MATH BOOTCAMP FOR GRADES K-1 TO K-4

KALMAN TOTH M.A., M. PHIL.

MATH BOOTCAMP FOR GRADES K-1 TO K-4

COPYRIGHT © 2013 BY KALMAN TOTH

Third Edition

ISBN-13: 978-1491078303

ISBN-10: 1491078308

TRADEMARK NOTICES

WARNING AND DISCLAIMER

CONTENTS AT A GLANCE

ABOUT THE AUTHOR

KALMAN TOTH, M.A. PHYSICS COLUMBIA UNIVERSITY & M.PHIL. IN COMPUTING SCIENCE COLUMBIA UNIVERSITY, IS AN SQL DATABASE DESIGN AND BUSINESS INTELLIGENCE SPECIALIST. HIS PROFESSIONAL RESEARCH INTEREST IS ARTIFICIAL INTELLIGENCE. HE IS CONVINCED THAT MACHINE INTELLIGENCE WILL NOT ONLY REPLACE HUMAN INTELLIGENCE BUT SURPASS IT MILLION TIMES IN THE NEAR FUTURE. HIS HOBBY IS FLYING GLIDERS, VINTAGE FIGHTER & BOMBER PLANES.

OTHER BOOKS BY THE AUTHOR:

Brain Puzzles for Adults

Brain Puzzles for Kids

IQ King Puzzles: Book #1

10000 Word Scramble Puzzles to Improve Your IQ

Lapple Puzzle Book #21: 1000 Overlapping Words Puzzles

400 Word Search Puzzles to Raise Your IQ

Brain Puzzles for Stroke Patients

1000 Diagonal Word Square Puzzles to Improve Your IQ

1000 SUDOKU Puzzles to Improve Your IQ

English Prodigy Spelling Bootcamp for Grades 7 & 8

Cognitive English Spelling Bootcamp for High School & College

Math Bootcamp for Grades K-5 to K-8

SQL Server 2012 Programming

SQL Server 2012 Database Design

CONTENTS

INTRODUCTION

BASIC MATH - ADDITION, SUBTRACTION, MULTIPLICATION & DIVISION WITH/WITHOUT COMPOSITION - IS THE FOUNDATION FOR BUILDING SOLID MATHEMATICAL SKILL. THIS BOOK IS DEDICATED TO THAT SINGLE GOAL: DRILLS AT FOUR GRADE LEVELS.

QUOTE: "WHY IS IT SO IMPORTANT FOR CHILDREN TO MEMORIZE MATH FACTS IN ORDER TO SUCCEED ACADEMICALLY? QUITE SIMPLY, A LACK OF FLUENCY IN BASIC MATH FACT RECALL SIGNIFICANTLY HINDERS A CHILD'S SUBSEQUENT PROGRESS WITH PROBLEM-SOLVING, ALGEBRA AND HIGHER-ORDER MATH CONCEPTS. THIS CAN HAVE A SERIOUS IMPACT ON A CHILD'S OVERALL SELF CONFIDENCE AND GENERAL ACADEMIC PERFORMANCE.

THERE HAS BEEN CONTROVERSY ABOUT THE NEED TO MEMORIZE MATH FACTS SINCE THE INTRODUCTION OF SIGNIFICANT REFORMS IN MATH CURRICULUM IN THE 1990S, WHICH LARGELY REPLACED ROTE MEMORIZATION WITH A NEW EMPHASIS ON INTEGRATIVE MATH TEACHING. THIS INVOLVES TEACHING MANY DIFFERENT CONCEPTS AT THE SAME TIME INSTEAD OF SEQUENTIALLY, AND USING MANIPULATIVES IN PLACE OF NUMBERS TO ILLUSTRATE MATHEMATICAL CONCEPTS LONG AFTER NUMBER SENSE SHOULD HAVE BEEN MASTERED. LEADING RESEARCHERS HAVE CAUTIONED THAT THIS HAS RESULTED IN A MATH CURRICULUM THAT IS TOO COMPLEX IN THE EARLY GRADES, INTRODUCING ADVANCED MATHEMATICAL CONCEPTS BEFORE CHILDREN HAVE MASTERED BASIC COMPUTATION.

A REPORT BY TOM LOVELESS, DIRECTOR OF THE BROWN CENTER ON EDUCATION POLICY AT THE BROOKINGS INSTITUTION, WASHINGTON DC, STATES THAT "YOUNGSTERS WHO HAVE NOT MASTERED WHOLE NUMBER ARITHMETIC BY THE END OF 4TH GRADE ARE AT RISK OF LATER BECOMING REMEDIAL STUDENTS IN MATHEMATICS" AND URGES THAT EVERY STUDENT IN THE NATION SHOULD RECEIVE A THOROUGH GROUNDING IN ARITHMETIC."

Article: **The Benefits of Memorizing Math Facts**   http://bit.ly/bRMcoE By Margaret Groves, M.Phil., M.Ed. ELS http://quickreckoning.com/math_research.htm

QUOTE: "A FRIEND OF MINE TEACHES REMEDIAL MATH AT THE COMMUNITY COLLEGE LEVEL. WE WERE DISCUSSING THE PROBLEM OF A NUMBER OF STUDENTS WHO NEVER SEEM TO HAVE THEIR ADDITION FACTS MASTERED (MUCH LESS THEIR MULTIPLICATION FACTS). HE WROTE:

"I REMEMBER AS A YOUNG MATH TEACHER WONDERING HOW MANY HOURS OF FLASH CARD DRILL IT TAKES IN THE ELEMENTARY GRADES TO BECOME FLUENT IN THE ADDITION AND MULTIPLICATION FACTS. I COULD IMAGINE TEN MINUTES A DAY OF ACTUAL FLASH CARD DRILL, FIVE DAYS A WEEK, FOR 45 WEEKS IN ONE GRADE, A TOTAL OF 50 HOURS IF I MULTIPLIED CORRECTLY, MIGHT BE A REASONABLE GUESS. SURELY IT HAS BEEN STUDIED. WELL, IF IT HAS BEEN STUDIED I HAVE NEVER SEEN ANY EVIDENCE OF IT IN THE LAST FIFTY YEARS. I THOUGHT OF THAT AS PROBABLY PRETTY BASIC KNOWLEDGE ABOUT THE TEACHING OF ARITHMETIC."

I'D ESTIMATE THAT AT LEAST 30 HOURS OF DRILL, SPREAD OVER A PERIOD OF TIME WOULD BE REQUIRED FOR AN AVERAGE CHILD TO LEARN ADDITION FACTS, AND AN EQUAL AMOUNT OF TIME LATER ON TO LEARN MULTIPLICATION FACTS. BUT THERE IS NO CLASSROOM TIME PROVIDED FOR THIS IN THE MATH CURRICULUM.

THE PROBLEM HERE, SPEAKING AS A THIRD-GRADE TEACHER OF 8-AND-9-YEAR-OLD STUDENTS FOR A DECADE, IS THAT THE ELEMENTARY MATH CURRICULUM (IN AMERICA) IS NOT STRUCTURED TO PROVIDE ANY TIME FOR DRILL SUCH AS HE DESCRIBES. EVEN WHEN I WAS A CHILD IN THE EARLY 1960'S, WE DID NOT HAVE DRILL OF THAT TYPE IN ELEMENTARY SCHOOL. MY MOTHER WORKED ON FLASH CARDS WITH ME 10-15 MINUTES EVERY DAY BEFORE I WAS ALLOWED TO PLAY. I HATED EVERY MOMENT OF IT, BUT SAW THE VALUE OF IT WHEN I GOT INTO THE WORKING WORLD IN MY 20'S AND WAS USING MULTIPLICATION EVERY DAY, KNOWING MY MULTIPLICATION TABLES BY HEART THANKS TO HER EFFORTS."

Article:
Why So Many Elementary Students Aren't Mastering Basic Math Facts   http://bit.ly/nq5zhw
http://expattutor.wordpress.com/2011/07/14/why-so-many-elementary-students-arent-mastering-basic-math-facts/

WHO THIS BOOK IS FOR

THIS BOOK WITH BASIC MATH DRILLS & SHOPPING MATH IS FOR THE STUDENT, PARENT, TEACHER OR TUTOR. THE EXERCISES REVOLVE AROUND DIRECT OR COMPOSITION ADDITION, SUBTRACTION, MULTIPLICATION & DIVISION WHICH FORMS THE VERY BASIS OF OUR MATHEMATICAL THINKING AND NUMERICAL INTELLIGENCE. THE NUMBER & WORD MATH & COMPOSITION EXERCISES ARE AT 4 DIFFERENT LEVELS ALLOWING CUSTOMIZATION FOR EACH STUDENT.

ABOUT THIS BOOK

MATH DRILLS ARE ORGANIZED THE FOLLOWING WAY:

> - MATH OPERATION WITHOUT RESULT
> - MATH OPERATION WITH RESULT
> - LEFT OPERAND MISSING (COMPOSITION)
> - RIGHT OPERAND MISSING (COMPOSITION)
> - VERBAL SHOPPING MATH

THIS BOOK CAN BE USED AS IS OR PAGES CUT OUT PLACED INTO SEE-THROUGH PAGE PROTECTORS (AVAILABLE IN OFFICE PRODUCTS STORES) AND ASSEMBLED IN 3-RING BINDERS. USE A RULER AND A UTILITY KNIFE TO CUT OUT A PAGE.

"APPARATUS INTELLIGENTIA

VINCET

HUMANUM INTELLIGENTIA"

DEDICATED TO

JOHN VON NEUMANN

Father of the Computer

THIS PAGE IS INTENTIONALLY LEFT BLANK.

CHAPTER 1: K-1 MATH EXERCISES

ADDITIONS

1.	11 + 4 = _____
2.	17 + 15 = _____
3.	5 + 7 = _____
4.	4 + 5 = _____
5.	20 + 18 = _____
6.	5 + 20 = _____
7.	14 + 7 = _____
8.	2 + 4 = _____
9.	7 + 18 = _____
10.	14 + 16 = _____
11.	19 + 15 = _____
12.	17 + 12 = _____
13.	18 + 5 = _____
14.	11 + 19 = _____

15.	12 + 18 = _____
16.	6 + 7 = _____
17.	16 + 20 = _____
18.	12 + 16 = _____
19.	3 + 12 = _____
20.	6 + 15 = _____
21.	5 + 18 = _____
22.	18 + 16 = _____
23.	19 + 19 = _____
24.	10 + 4 = _____
25.	2 + 2 = _____
26.	5 + 12 = _____
27.	11 + 11 = _____
28.	3 + 12 = _____
29.	7 + 13 = _____
30.	16 + 6 = _____
31.	20 + 9 = _____

32.   6 + 18 = ____

33.   14 + 2 = ____

34.   7 + 7 = ____

35.   9 + 6 = ____

36.   11 + 10 = ____

37.   12 + 5 = ____

38.   7 + 18 = ____

39.   15 + 12 = ____

40.   10 + 2 = ____

41.   4 + 11 = ____

42.   17 + 7 = ____

43.   5 + 6 = ____

44.   10 + 2 = ____

45.   4 + 7 = ____

46.   15 + 11 = ____

47.   10 + 9 = ____

48.   13 + 14 = ____

49.	17 + 13 = _____
50.	8 + 7 = _____
51.	20 + 3 = _____
52.	17 + 6 = _____
53.	9 + 20 = _____
54.	5 + 18 = _____
55.	8 + 7 = _____
56.	13 + 11 = _____
57.	17 + 4 = _____
58.	3 + 6 = _____
59.	8 + 13 = _____
60.	19 + 18 = _____
61.	13 + 10 = _____
62.	10 + 13 = _____
63.	11 + 4 = _____
64.	9 + 19 = _____
65.	2 + 15 = _____

66.	14 + 12 = _____
67.	10 + 4 = _____
68.	17 + 19 = _____
69.	10 + 20 = _____
70.	8 + 7 = _____
71.	19 + 11 = _____
72.	20 + 19 = _____
73.	14 + 13 = _____
74.	7 + 12 = _____
75.	19 + 19 = _____
76.	15 + 3 = _____
77.	8 + 4 = _____
78.	17 + 15 = _____
79.	19 + 12 = _____
80.	11 + 13 = _____
81.	8 + 4 = _____
82.	12 + 10 = _____

83.	14 + 19 = ____
84.	12 + 7 = ____
85.	2 + 4 = ____
86.	5 + 3 = ____
87.	11 + 18 = ____
88.	18 + 15 = ____
89.	5 + 17 = ____
90.	6 + 4 = ____
91.	17 + 16 = ____
92.	12 + 13 = ____
93.	17 + 12 = ____
94.	6 + 19 = ____
95.	8 + 10 = ____
96.	20 + 20 = ____
97.	15 + 4 = ____
98.	6 + 15 = ____
99.	4 + 9 = ____

100.	7 + 3 = _____
1.	11 + 4 = 15
2.	17 + 15 = 32
3.	5 + 7 = 12
4.	4 + 5 = 9
5.	20 + 18 = 38
6.	5 + 20 = 25
7.	14 + 7 = 21
8.	2 + 4 = 6
9.	7 + 18 = 25
10.	14 + 16 = 30
11.	19 + 15 = 34
12.	17 + 12 = 29
13.	18 + 5 = 23
14.	11 + 19 = 30
15.	12 + 18 = 30
16.	6 + 7 = 13

17.	16 + 20 = 36
18.	12 + 16 = 28
19.	3 + 12 = 15
20.	6 + 15 = 21
21.	5 + 18 = 23
22.	18 + 16 = 34
23.	19 + 19 = 38
24.	10 + 4 = 14
25.	2 + 2 = 4
26.	5 + 12 = 17
27.	11 + 11 = 22
28.	3 + 12 = 15
29.	7 + 13 = 20
30.	16 + 6 = 22
31.	20 + 9 = 29
32.	6 + 18 = 24
33.	14 + 2 = 16

34.	7 + 7 = 14
35.	9 + 6 = 15
36.	11 + 10 = 21
37.	12 + 5 = 17
38.	7 + 18 = 25
39.	15 + 12 = 27
40.	10 + 2 = 12
41.	4 + 11 = 15
42.	17 + 7 = 24
43.	5 + 6 = 11
44.	10 + 2 = 12
45.	4 + 7 = 11
46.	15 + 11 = 26
47.	10 + 9 = 19
48.	13 + 14 = 27
49.	17 + 13 = 30
50.	8 + 7 = 15

51.	20 + 3 = 23
52.	17 + 6 = 23
53.	9 + 20 = 29
54.	5 + 18 = 23
55.	8 + 7 = 15
56.	13 + 11 = 24
57.	17 + 4 = 21
58.	3 + 6 = 9
59.	8 + 13 = 21
60.	19 + 18 = 37
61.	13 + 10 = 23
62.	10 + 13 = 23
63.	11 + 4 = 15
64.	9 + 19 = 28
65.	2 + 15 = 17
66.	14 + 12 = 26
67.	10 + 4 = 14

68.	17 + 19 = 36
69.	10 + 20 = 30
70.	8 + 7 = 15
71.	19 + 11 = 30
72.	20 + 19 = 39
73.	14 + 13 = 27
74.	7 + 12 = 19
75.	19 + 19 = 38
76.	15 + 3 = 18
77.	8 + 4 = 12
78.	17 + 15 = 32
79.	19 + 12 = 31
80.	11 + 13 = 24
81.	8 + 4 = 12
82.	12 + 10 = 22
83.	14 + 19 = 33
84.	12 + 7 = 19

85.	2 + 4 = 6
86.	5 + 3 = 8
87.	11 + 18 = 29
88.	18 + 15 = 33
89.	5 + 17 = 22
90.	6 + 4 = 10
91.	17 + 16 = 33
92.	12 + 13 = 25
93.	17 + 12 = 29
94.	6 + 19 = 25
95.	8 + 10 = 18
96.	20 + 20 = 40
97.	15 + 4 = 19
98.	6 + 15 = 21
99.	4 + 9 = 13
100.	7 + 3 = 10
1.	____ + 4 = 15

2.	____ + 15 = 32
3.	____ + 7 = 12
4.	____ + 5 = 9
5.	____ + 18 = 38
6.	____ + 20 = 25
7.	____ + 7 = 21
8.	____ + 4 = 6
9.	____ + 18 = 25
10.	____ + 16 = 30
11.	____ + 15 = 34
12.	____ + 12 = 29
13.	____ + 5 = 23
14.	____ + 19 = 30
15.	____ + 18 = 30
16.	____ + 7 = 13
17.	____ + 20 = 36
18.	____ + 16 = 28

19.	____ + 12 = 15
20.	____ + 15 = 21
21.	____ + 18 = 23
22.	____ + 16 = 34
23.	____ + 19 = 38
24.	____ + 4 = 14
25.	____ + 2 = 4
26.	____ + 12 = 17
27.	____ + 11 = 22
28.	____ + 12 = 15
29.	____ + 13 = 20
30.	____ + 6 = 22
31.	____ + 9 = 29
32.	____ + 18 = 24
33.	____ + 2 = 16
34.	____ + 7 = 14
35.	____ + 6 = 15

36.	_____ + 10 = 21
37.	_____ + 5 = 17
38.	_____ + 18 = 25
39.	_____ + 12 = 27
40.	_____ + 2 = 12
41.	_____ + 11 = 15
42.	_____ + 7 = 24
43.	_____ + 6 = 11
44.	_____ + 2 = 12
45.	_____ + 7 = 11
46.	_____ + 11 = 26
47.	_____ + 9 = 19
48.	_____ + 14 = 27
49.	_____ + 13 = 30
50.	_____ + 7 = 15
51.	_____ + 3 = 23
52.	_____ + 6 = 23

53.	_____ + 20 = 29
54.	_____ + 18 = 23
55.	_____ + 7 = 15
56.	_____ + 11 = 24
57.	_____ + 4 = 21
58.	_____ + 6 = 9
59.	_____ + 13 = 21
60.	_____ + 18 = 37
61.	_____ + 10 = 23
62.	_____ + 13 = 23
63.	_____ + 4 = 15
64.	_____ + 19 = 28
65.	_____ + 15 = 17
66.	_____ + 12 = 26
67.	_____ + 4 = 14
68.	_____ + 19 = 36
69.	_____ + 20 = 30

70.	____ + 7 = 15
71.	____ + 11 = 30
72.	____ + 19 = 39
73.	____ + 13 = 27
74.	____ + 12 = 19
75.	____ + 19 = 38
76.	____ + 3 = 18
77.	____ + 4 = 12
78.	____ + 15 = 32
79.	____ + 12 = 31
80.	____ + 13 = 24
81.	____ + 4 = 12
82.	____ + 10 = 22
83.	____ + 19 = 33
84.	____ + 7 = 19
85.	____ + 4 = 6
86.	____ + 3 = 8

87.	____ + 18 = 29
88.	____ + 15 = 33
89.	____ + 17 = 22
90.	____ + 4 = 10
91.	____ + 16 = 33
92.	____ + 13 = 25
93.	____ + 12 = 29
94.	____ + 19 = 25
95.	____ + 10 = 18
96.	____ + 20 = 40
97.	____ + 4 = 19
98.	____ + 15 = 21
99.	____ + 9 = 13
100.	____ + 3 = 10
1.	11 + ____ = 15
2.	17 + ____ = 32
3.	5 + ____ = 12

4.	4 + ____ = 9
5.	20 + ____ = 38
6.	5 + ____ = 25
7.	14 + ____ = 21
8.	2 + ____ = 6
9.	7 + ____ = 25
10.	14 + ____ = 30
11.	19 + ____ = 34
12.	17 + ____ = 29
13.	18 + ____ = 23
14.	11 + ____ = 30
15.	12 + ____ = 30
16.	6 + ____ = 13
17.	16 + ____ = 36
18.	12 + ____ = 28
19.	3 + ____ = 15
20.	6 + ____ = 21

21.	$5 + \rule{1cm}{0.1mm} = 23$
22.	$18 + \rule{1cm}{0.1mm} = 34$
23.	$19 + \rule{1cm}{0.1mm} = 38$
24.	$10 + \rule{1cm}{0.1mm} = 14$
25.	$2 + \rule{1cm}{0.1mm} = 4$
26.	$5 + \rule{1cm}{0.1mm} = 17$
27.	$11 + \rule{1cm}{0.1mm} = 22$
28.	$3 + \rule{1cm}{0.1mm} = 15$
29.	$7 + \rule{1cm}{0.1mm} = 20$
30.	$16 + \rule{1cm}{0.1mm} = 22$
31.	$20 + \rule{1cm}{0.1mm} = 29$
32.	$6 + \rule{1cm}{0.1mm} = 24$
33.	$14 + \rule{1cm}{0.1mm} = 16$
34.	$7 + \rule{1cm}{0.1mm} = 14$
35.	$9 + \rule{1cm}{0.1mm} = 15$
36.	$11 + \rule{1cm}{0.1mm} = 21$
37.	$12 + \rule{1cm}{0.1mm} = 17$

38.	$7 + \underline{\quad} = 25$
39.	$15 + \underline{\quad} = 27$
40.	$10 + \underline{\quad} = 12$
41.	$4 + \underline{\quad} = 15$
42.	$17 + \underline{\quad} = 24$
43.	$5 + \underline{\quad} = 11$
44.	$10 + \underline{\quad} = 12$
45.	$4 + \underline{\quad} = 11$
46.	$15 + \underline{\quad} = 26$
47.	$10 + \underline{\quad} = 19$
48.	$13 + \underline{\quad} = 27$
49.	$17 + \underline{\quad} = 30$
50.	$8 + \underline{\quad} = 15$
51.	$20 + \underline{\quad} = 23$
52.	$17 + \underline{\quad} = 23$
53.	$9 + \underline{\quad} = 29$
54.	$5 + \underline{\quad} = 23$

55.	$8 + \underline{\hspace{1cm}} = 15$
56.	$13 + \underline{\hspace{1cm}} = 24$
57.	$17 + \underline{\hspace{1cm}} = 21$
58.	$3 + \underline{\hspace{1cm}} = 9$
59.	$8 + \underline{\hspace{1cm}} = 21$
60.	$19 + \underline{\hspace{1cm}} = 37$
61.	$13 + \underline{\hspace{1cm}} = 23$
62.	$10 + \underline{\hspace{1cm}} = 23$
63.	$11 + \underline{\hspace{1cm}} = 15$
64.	$9 + \underline{\hspace{1cm}} = 28$
65.	$2 + \underline{\hspace{1cm}} = 17$
66.	$14 + \underline{\hspace{1cm}} = 26$
67.	$10 + \underline{\hspace{1cm}} = 14$
68.	$17 + \underline{\hspace{1cm}} = 36$
69.	$10 + \underline{\hspace{1cm}} = 30$
70.	$8 + \underline{\hspace{1cm}} = 15$
71.	$19 + \underline{\hspace{1cm}} = 30$

72.　　　$20 + \_\_\_\_ = 39$

73.　　　$14 + \_\_\_\_ = 27$

74.　　　$7 + \_\_\_\_ = 19$

75.　　　$19 + \_\_\_\_ = 38$

76.　　　$15 + \_\_\_\_ = 18$

77.　　　$8 + \_\_\_\_ = 12$

78.　　　$17 + \_\_\_\_ = 32$

79.　　　$19 + \_\_\_\_ = 31$

80.　　　$11 + \_\_\_\_ = 24$

81.　　　$8 + \_\_\_\_ = 12$

82.　　　$12 + \_\_\_\_ = 22$

83.　　　$14 + \_\_\_\_ = 33$

84.　　　$12 + \_\_\_\_ = 19$

85.　　　$2 + \_\_\_\_ = 6$

86.　　　$5 + \_\_\_\_ = 8$

87.　　　$11 + \_\_\_\_ = 29$

88.　　　$18 + \_\_\_\_ = 33$

89.	5 + ____ = 22
90.	6 + ____ = 10
91.	17 + ____ = 33
92.	12 + ____ = 25
93.	17 + ____ = 29
94.	6 + ____ = 25
95.	8 + ____ = 18
96.	20 + ____ = 40
97.	15 + ____ = 19
98.	6 + ____ = 21
99.	4 + ____ = 13
100.	7 + ____ = 10
1.	Eleven + Four = ____
2.	Seventeen + Fifteen = ____
3.	Five + Seven = ____
4.	Four + Five = ____
5.	Twenty + Eighteen = ____

6.	Five + Twenty = _____
7.	Fourteen + Seven = _____
8.	Two + Four = _____
9.	Seven + Eighteen = _____
10.	Fourteen + Sixteen = _____
11.	Nineteen + Fifteen = _____
12.	Seventeen + Twelve = _____
13.	Eighteen + Five = _____
14.	Eleven + Nineteen = _____
15.	Twelve + Eighteen = _____
16.	Six + Seven = _____
17.	Sixteen + Twenty = _____
18.	Twelve + Sixteen = _____
19.	Three + Twelve = _____
20.	Six + Fifteen = _____
21.	Five + Eighteen = _____
22.	Eighteen + Sixteen = _____

23.	Nineteen + Nineteen = _____
24.	Ten + Four = _____
25.	Two + Two = _____
26.	Five + Twelve = _____
27.	Eleven + Eleven = _____
28.	Three + Twelve = _____
29.	Seven + Thirteen = _____
30.	Sixteen + Six = _____
1.	Eleven + Four = Fifteen
2.	Seventeen + Fifteen = Thirty-Two
3.	Five + Seven = Twelve
4.	Four + Five = Nine
5.	Twenty + Eighteen = Thirty-Eight
6.	Five + Twenty = Twenty-Five
7.	Fourteen + Seven = Twenty-One
8.	Two + Four = Six
9.	Seven + Eighteen = Twenty-Five

10.	Fourteen + Sixteen = Thirty
11.	Nineteen + Fifteen = Thirty-Four
12.	Seventeen + Twelve = Twenty-Nine
13.	Eighteen + Five = Twenty-Three
14.	Eleven + Nineteen = Thirty
15.	Twelve + Eighteen = Thirty
16.	Six + Seven = Thirteen
17.	Sixteen + Twenty = Thirty-Six
18.	Twelve + Sixteen = Twenty-Eight
19.	Three + Twelve = Fifteen
20.	Six + Fifteen = Twenty-One
21.	Five + Eighteen = Twenty-Three
22.	Eighteen + Sixteen = Thirty-Four
23.	Nineteen + Nineteen = Thirty-Eight
24.	Ten + Four = Fourteen
25.	Two + Two = Four

CHAPTER 1: K-1 MATH EXERCISES

26.	Five + Twelve = Seventeen
27.	Eleven + Eleven = Twenty-Two
28.	Three + Twelve = Fifteen
29.	Seven + Thirteen = Twenty
30.	Sixteen + Six = Twenty-Two
1.	_____ + Four = Fifteen
2.	_____ + Fifteen = Thirty-Two
3.	_____ + Seven = Twelve
4.	_____ + Five = Nine
5.	_____ + Eighteen = Thirty-Eight
6.	_____ + Twenty = Twenty-Five
7.	_____ + Seven = Twenty-One
8.	_____ + Four = Six
9.	_____ + Eighteen = Twenty-Five
10.	_____ + Sixteen = Thirty
11.	_____ + Fifteen = Thirty-Four
12.	_____ + Twelve = Twenty-Nine

13.	_____ + Five = Twenty-Three
14.	_____ + Nineteen = Thirty
15.	_____ + Eighteen = Thirty
16.	_____ + Seven = Thirteen
17.	_____ + Twenty = Thirty-Six
18.	_____ + Sixteen = Twenty-Eight
19.	_____ + Twelve = Fifteen
20.	_____ + Fifteen = Twenty-One
21.	_____ + Eighteen = Twenty-Three
22.	_____ + Sixteen = Thirty-Four
23.	_____ + Nineteen = Thirty-Eight
24.	_____ + Four = Fourteen
25.	_____ + Two = Four
26.	_____ + Twelve = Seventeen
27.	_____ + Eleven = Twenty-Two
28.	_____ + Twelve = Fifteen
29.	_____ + Thirteen = Twenty

30.	_____ + Six = Twenty-Two
1.	Eleven + _____ = Fifteen
2.	Seventeen + _____ = Thirty-Two
3.	Five + _____ = Twelve
4.	Four + _____ = Nine
5.	Twenty + _____ = Thirty-Eight
6.	Five + _____ = Twenty-Five
7.	Fourteen + _____ = Twenty-One
8.	Two + _____ = Six
9.	Seven + _____ = Twenty-Five
10.	Fourteen + _____ = Thirty
11.	Nineteen + _____ = Thirty-Four
12.	Seventeen + _____ = Twenty-Nine
13.	Eighteen + _____ = Twenty-Three
14.	Eleven + _____ = Thirty
15.	Twelve + _____ = Thirty
16.	Six + _____ = Thirteen

17.	Sixteen + _____ = Thirty-Six
18.	Twelve + _____ = Twenty-Eight
19.	Three + _____ = Fifteen
20.	Six + _____ = Twenty-One
21.	Five + _____ = Twenty-Three
22.	Eighteen + _____ = Thirty-Four
23.	Nineteen + _____ = Thirty-Eight
24.	Ten + _____ = Fourteen
25.	Two + _____ = Four
26.	Five + _____ = Seventeen
27.	Eleven + _____ = Twenty-Two
28.	Three + _____ = Fifteen

HIGHWAY TRAFFIC ON INTERSTATE 78 IN PENNSYLVANIA

CHAPTER 1: K-1 MATH EXERCISES

SUBTRACTIONS

1.	13 - 4 = _____
2.	17 - 15 = _____
3.	18 - 11 = _____
4.	17 - 15 = _____
5.	20 - 12 = _____
6.	13 - 9 = _____
7.	14 - 3 = _____
8.	19 - 17 = _____
9.	18 - 11 = _____
10.	13 - 10 = _____
11.	19 - 14 = _____
12.	14 - 9 = _____
13.	13 - 11 = _____
14.	20 - 14 = _____
15.	18 - 3 = _____

16.	16 - 4 = _____
17.	19 - 8 = _____
18.	20 - 15 = _____
19.	18 - 5 = _____
20.	17 - 6 = _____
21.	19 - 8 = _____
22.	6 - 4 = _____
23.	8 - 7 = _____
24.	20 - 6 = _____
25.	12 - 11 = _____
26.	9 - 5 = _____
27.	14 - 7 = _____
28.	12 - 11 = _____
29.	15 - 4 = _____
30.	4 - 3 = _____
31.	19 - 6 = _____
32.	4 - 2 = _____

33. 16 - 7 = ____

34. 18 - 10 = ____

35. 11 - 3 = ____

36. 9 - 4 = ____

37. 10 - 9 = ____

38. 15 - 5 = ____

39. 19 - 15 = ____

40. 10 - 8 = ____

41. 15 - 13 = ____

42. 9 - 4 = ____

43. 16 - 12 = ____

44. 5 - 3 = ____

45. 12 - 5 = ____

46. 10 - 8 = ____

47. 4 - 2 = ____

48. 11 - 7 = ____

49. 16 - 11 = ____

50.   10 - 5 = _____

51.   5 - 2 = _____

52.   20 - 15 = _____

53.   6 - 4 = _____

54.   18 - 6 = _____

55.   20 - 2 = _____

56.   16 - 8 = _____

57.   19 - 12 = _____

58.   16 - 9 = _____

59.   17 - 2 = _____

60.   18 - 6 = _____

61.   11 - 3 = _____

62.   16 - 13 = _____

63.   17 - 11 = _____

64.   15 - 14 = _____

65.   20 - 10 = _____

66.   7 - 2 = _____

67.	14 - 8 = _____
68.	11 - 7 = _____
69.	20 - 12 = _____
70.	17 - 9 = _____
71.	4 - 3 = _____
72.	10 - 4 = _____
73.	18 - 3 = _____
74.	12 - 5 = _____
75.	13 - 9 = _____
76.	20 - 5 = _____
77.	17 - 11 = _____
78.	19 - 8 = _____
79.	15 - 10 = _____
80.	20 - 4 = _____
81.	9 - 8 = _____
82.	6 - 4 = _____
83.	20 - 14 = _____

84.	7 - 3 = ____
85.	13 - 7 = ____
86.	18 - 10 = ____
87.	13 - 2 = ____
88.	19 - 13 = ____
89.	17 - 4 = ____
90.	10 - 6 = ____
91.	19 - 15 = ____
92.	13 - 9 = ____
93.	3 - 2 = ____
94.	13 - 4 = ____
95.	14 - 3 = ____
96.	11 - 5 = ____
97.	10 - 3 = ____
98.	16 - 12 = ____
99.	18 - 9 = ____
100.	19 - 17 = ____

1.	13 - 4 = 9
2.	17 - 15 = 2
3.	18 - 11 = 7
4.	17 - 15 = 2
5.	20 - 12 = 8
6.	13 - 9 = 4
7.	14 - 3 = 11
8.	19 - 17 = 2
9.	18 - 11 = 7
10.	13 - 10 = 3
11.	19 - 14 = 5
12.	14 - 9 = 5
13.	13 - 11 = 2
14.	20 - 14 = 6
15.	18 - 3 = 15
16.	16 - 4 = 12
17.	19 - 8 = 11

18.	20 - 15 = 5
19.	18 - 5 = 13
20.	17 - 6 = 11
21.	19 - 8 = 11
22.	6 - 4 = 2
23.	8 - 7 = 1
24.	20 - 6 = 14
25.	12 - 11 = 1
26.	9 - 5 = 4
27.	14 - 7 = 7
28.	12 - 11 = 1
29.	15 - 4 = 11
30.	4 - 3 = 1
31.	19 - 6 = 13
32.	4 - 2 = 2
33.	16 - 7 = 9
34.	18 - 10 = 8

35.	11 - 3 = 8
36.	9 - 4 = 5
37.	10 - 9 = 1
38.	15 - 5 = 10
39.	19 - 15 = 4
40.	10 - 8 = 2
41.	15 - 13 = 2
42.	9 - 4 = 5
43.	16 - 12 = 4
44.	5 - 3 = 2
45.	12 - 5 = 7
46.	10 - 8 = 2
47.	4 - 2 = 2
48.	11 - 7 = 4
49.	16 - 11 = 5
50.	10 - 5 = 5
51.	5 - 2 = 3

52.	20 - 15 = 5
53.	6 - 4 = 2
54.	18 - 6 = 12
55.	20 - 2 = 18
56.	16 - 8 = 8
57.	19 - 12 = 7
58.	16 - 9 = 7
59.	17 - 2 = 15
60.	18 - 6 = 12
61.	11 - 3 = 8
62.	16 - 13 = 3
63.	17 - 11 = 6
64.	15 - 14 = 1
65.	20 - 10 = 10
66.	7 - 2 = 5
67.	14 - 8 = 6
68.	11 - 7 = 4

69.	20 - 12 = 8
70.	17 - 9 = 8
71.	4 - 3 = 1
72.	10 - 4 = 6
73.	18 - 3 = 15
74.	12 - 5 = 7
75.	13 - 9 = 4
76.	20 - 5 = 15
77.	17 - 11 = 6
78.	19 - 8 = 11
79.	15 - 10 = 5
80.	20 - 4 = 16
81.	9 - 8 = 1
82.	6 - 4 = 2
83.	20 - 14 = 6
84.	7 - 3 = 4
85.	13 - 7 = 6

86.	18 - 10 = 8
87.	13 - 2 = 11
88.	19 - 13 = 6
89.	17 - 4 = 13
90.	10 - 6 = 4
91.	19 - 15 = 4
92.	13 - 9 = 4
93.	3 - 2 = 1
94.	13 - 4 = 9
95.	14 - 3 = 11
96.	11 - 5 = 6
97.	10 - 3 = 7
98.	16 - 12 = 4
99.	18 - 9 = 9
100.	19 - 17 = 2
1.	_____ - 4 = 9
2.	_____ - 15 = 2

3.   ____ - 11 = 7

4.   ____ - 15 = 2

5.   ____ - 12 = 8

6.   ____ - 9 = 4

7.   ____ - 3 = 11

8.   ____ - 17 = 2

9.   ____ - 11 = 7

10.   ____ - 10 = 3

11.   ____ - 14 = 5

12.   ____ - 9 = 5

13.   ____ - 11 = 2

14.   ____ - 14 = 6

15.   ____ - 3 = 15

16.   ____ - 4 = 12

17.   ____ - 8 = 11

18.   ____ - 15 = 5

19.   ____ - 5 = 13

20.   ____ - 6 = 11

21.   ____ - 8 = 11

22.   ____ - 4 = 2

23.   ____ - 7 = 1

24.   ____ - 6 = 14

25.   ____ - 11 = 1

26.   ____ - 5 = 4

27.   ____ - 7 = 7

28.   ____ - 11 = 1

29.   ____ - 4 = 11

30.   ____ - 3 = 1

31.   ____ - 6 = 13

32.   ____ - 2 = 2

33.   ____ - 7 = 9

34.   ____ - 10 = 8

35.   ____ - 3 = 8

36.   ____ - 4 = 5

37. ____ - 9 = 1

38. ____ - 5 = 10

39. ____ - 15 = 4

40. ____ - 8 = 2

41. ____ - 13 = 2

42. ____ - 4 = 5

43. ____ - 12 = 4

44. ____ - 3 = 2

45. ____ - 5 = 7

46. ____ - 8 = 2

47. ____ - 2 = 2

48. ____ - 7 = 4

49. ____ - 11 = 5

50. ____ - 5 = 5

51. ____ - 2 = 3

52. ____ - 15 = 5

53. ____ - 4 = 2

54.	_____ - 6 = 12
55.	_____ - 2 = 18
56.	_____ - 8 = 8
57.	_____ - 12 = 7
58.	_____ - 9 = 7
59.	_____ - 2 = 15
60.	_____ - 6 = 12
61.	_____ - 3 = 8
62.	_____ - 13 = 3
63.	_____ - 11 = 6
64.	_____ - 14 = 1
65.	_____ - 10 = 10
66.	_____ - 2 = 5
67.	_____ - 8 = 6
68.	_____ - 7 = 4
69.	_____ - 12 = 8
70.	_____ - 9 = 8

71.	_____ - 3 = 1
72.	_____ - 4 = 6
73.	_____ - 3 = 15
74.	_____ - 5 = 7
75.	_____ - 9 = 4
76.	_____ - 5 = 15
77.	_____ - 11 = 6
78.	_____ - 8 = 11
79.	_____ - 10 = 5
80.	_____ - 4 = 16
81.	_____ - 8 = 1
82.	_____ - 4 = 2
83.	_____ - 14 = 6
84.	_____ - 3 = 4
85.	_____ - 7 = 6
86.	_____ - 10 = 8
87.	_____ - 2 = 11

88.	____ - 13 = 6
89.	____ - 4 = 13
90.	____ - 6 = 4
91.	____ - 15 = 4
92.	____ - 9 = 4
93.	____ - 2 = 1
94.	____ - 4 = 9
95.	____ - 3 = 11
96.	____ - 5 = 6
97.	____ - 3 = 7
98.	____ - 12 = 4
99.	____ - 9 = 9
100.	____ - 17 = 2
1.	13 - ____ = 9
2.	17 - ____ = 2
3.	18 - ____ = 7
4.	17 - ____ = 2

5.	20 - ____ = 8
6.	13 - ____ = 4
7.	14 - ____ = 11
8.	19 - ____ = 2
9.	18 - ____ = 7
10.	13 - ____ = 3
11.	19 - ____ = 5
12.	14 - ____ = 5
13.	13 - ____ = 2
14.	20 - ____ = 6
15.	18 - ____ = 15
16.	16 - ____ = 12
17.	19 - ____ = 11
18.	20 - ____ = 5
19.	18 - ____ = 13
20.	17 - ____ = 11
21.	19 - ____ = 11

22.	6 - _____ = 2
23.	8 - _____ = 1
24.	20 - _____ = 14
25.	12 - _____ = 1
26.	9 - _____ = 4
27.	14 - _____ = 7
28.	12 - _____ = 1
29.	15 - _____ = 11
30.	4 - _____ = 1
31.	19 - _____ = 13
32.	4 - _____ = 2
33.	16 - _____ = 9
34.	18 - _____ = 8
35.	11 - _____ = 8
36.	9 - _____ = 5
37.	10 - _____ = 1
38.	15 - _____ = 10

39.	19 - ____ = 4
40.	10 - ____ = 2
41.	15 - ____ = 2
42.	9 - ____ = 5
43.	16 - ____ = 4
44.	5 - ____ = 2
45.	12 - ____ = 7
46.	10 - ____ = 2
47.	4 - ____ = 2
48.	11 - ____ = 4
49.	16 - ____ = 5
50.	10 - ____ = 5
51.	5 - ____ = 3
52.	20 - ____ = 5
53.	6 - ____ = 2
54.	18 - ____ = 12
55.	20 - ____ = 18

56.	$16 - \underline{\hspace{1cm}} = 8$
57.	$19 - \underline{\hspace{1cm}} = 7$
58.	$16 - \underline{\hspace{1cm}} = 7$
59.	$17 - \underline{\hspace{1cm}} = 15$
60.	$18 - \underline{\hspace{1cm}} = 12$
61.	$11 - \underline{\hspace{1cm}} = 8$
62.	$16 - \underline{\hspace{1cm}} = 3$
63.	$17 - \underline{\hspace{1cm}} = 6$
64.	$15 - \underline{\hspace{1cm}} = 1$
65.	$20 - \underline{\hspace{1cm}} = 10$
66.	$7 - \underline{\hspace{1cm}} = 5$
67.	$14 - \underline{\hspace{1cm}} = 6$
68.	$11 - \underline{\hspace{1cm}} = 4$
69.	$20 - \underline{\hspace{1cm}} = 8$
70.	$17 - \underline{\hspace{1cm}} = 8$
71.	$4 - \underline{\hspace{1cm}} = 1$
72.	$10 - \underline{\hspace{1cm}} = 6$

73.	18 - _____ = 15
74.	12 - _____ = 7
75.	13 - _____ = 4
76.	20 - _____ = 15
77.	17 - _____ = 6
78.	19 - _____ = 11
79.	15 - _____ = 5
80.	20 - _____ = 16
81.	9 - _____ = 1
82.	6 - _____ = 2
83.	20 - _____ = 6
84.	7 - _____ = 4
85.	13 - _____ = 6
86.	18 - _____ = 8
87.	13 - _____ = 11
88.	19 - _____ = 6
89.	17 - _____ = 13

90.	10 - ____ = 4
91.	19 - ____ = 4
92.	13 - ____ = 4
93.	3 - ____ = 1
94.	13 - ____ = 9
95.	14 - ____ = 11
96.	11 - ____ = 6
97.	10 - ____ = 7
98.	16 - ____ = 4
99.	18 - ____ = 9
100.	19 - ____ = 2
1.	Thirteen - Four = ____
2.	Seventeen - Fifteen = ____
3.	Eighteen - Eleven = ____
4.	Seventeen - Fifteen = ____
5.	Twenty - Twelve = ____
6.	Thirteen - Nine = ____

7.	Fourteen - Three = _____
8.	Nineteen - Seventeen = _____
9.	Eighteen - Eleven = _____
10.	Thirteen - Ten = _____
11.	Nineteen - Fourteen = _____
12.	Fourteen - Nine = _____
13.	Thirteen - Eleven = _____
14.	Twenty - Fourteen = _____
15.	Eighteen - Three = _____
16.	Sixteen - Four = _____
17.	Nineteen - Eight = _____
18.	Twenty - Fifteen = _____
19.	Eighteen - Five = _____
20.	Seventeen - Six = _____
21.	Nineteen - Eight = _____
22.	Six - Four = _____
23.	Eight - Seven = _____

24.	Twenty - Six = ____
25.	Twelve - Eleven = ____
26.	Nine - Five = ____
27.	Fourteen - Seven = ____
28.	Twelve - Eleven = ____
29.	Fifteen - Four = ____
30.	Four - Three = ____
1.	Thirteen - Four = Nine
2.	Seventeen - Fifteen = Two
3.	Eighteen - Eleven = Seven
4.	Seventeen - Fifteen = Two
5.	Twenty - Twelve = Eight
6.	Thirteen - Nine = Four
7.	Fourteen - Three = Eleven
8.	Nineteen - Seventeen = Two
9.	Eighteen - Eleven = Seven
10.	Thirteen - Ten = Three

11.	Nineteen - Fourteen = Five
12.	Fourteen - Nine = Five
13.	Thirteen - Eleven = Two
14.	Twenty - Fourteen = Six
15.	Eighteen - Three = Fifteen
16.	Sixteen - Four = Twelve
17.	Nineteen - Eight = Eleven
18.	Twenty - Fifteen = Five
19.	Eighteen - Five = Thirteen
20.	Seventeen - Six = Eleven
21.	Nineteen - Eight = Eleven
22.	Six - Four = Two
23.	Eight - Seven = One
24.	Twenty - Six = Fourteen
25.	Twelve - Eleven = One
26.	Nine - Five = Four
27.	Fourteen - Seven = Seven

28.	Twelve - Eleven = One
29.	Fifteen - Four = Eleven
30.	Four - Three = One
1.	_____ - Four = Nine
2.	_____ - Fifteen = Two
3.	_____ - Eleven = Seven
4.	_____ - Fifteen = Two
5.	_____ - Twelve = Eight
6.	_____ - Nine = Four
7.	_____ - Three = Eleven
8.	_____ - Seventeen = Two
9.	_____ - Eleven = Seven
10.	_____ - Ten = Three
11.	_____ - Fourteen = Five
12.	_____ - Nine = Five
13.	_____ - Eleven = Two
14.	_____ - Fourteen = Six

15.	____ - Three = Fifteen
16.	____ - Four = Twelve
17.	____ - Eight = Eleven
18.	____ - Fifteen = Five
19.	____ - Five = Thirteen
20.	____ - Six = Eleven
21.	____ - Eight = Eleven
22.	____ - Four = Two
23.	____ - Seven = One
24.	____ - Six = Fourteen
25.	____ - Eleven = One
26.	____ - Five = Four
27.	____ - Seven = Seven
28.	____ - Eleven = One
29.	____ - Four = Eleven
30.	____ - Three = One
1.	Thirteen - ____ = Nine

2.   Seventeen - _____ = Two

3.   Eighteen - _____ = Seven

4.   Seventeen - _____ = Two

5.   Twenty - _____ = Eight

6.   Thirteen - _____ = Four

7.   Fourteen - _____ = Eleven

8.   Nineteen - _____ = Two

9.   Eighteen - _____ = Seven

10.   Thirteen - _____ = Three

11.   Nineteen - _____ = Five

12.   Fourteen - _____ = Five

13.   Thirteen - _____ = Two

14.   Twenty - _____ = Six

15.   Eighteen - _____ = Fifteen

16.   Sixteen - _____ = Twelve

17.   Nineteen - _____ = Eleven

18.   Twenty - _____ = Five

19.	**Eighteen - ____ = Thirteen**
20.	**Seventeen - ____ = Eleven**
21.	**Nineteen - ____ = Eleven**
22.	**Six - ____ = Two**
23.	**Eight - ____ = One**
24.	**Twenty - ____ = Fourteen**
25.	**Twelve - ____ = One**
26.	**Nine - ____ = Four**
27.	**Fourteen - ____ = Seven**
28.	**Twelve - ____ = One**
29.	**Fifteen - ____ = Eleven**
30.	**Four - ____ = One**

AMISH HORSE & CARRIAGE IN LANCASTER COUNTY PENNSYLVANIA

AMISH FARMER WITH SIX HORSE TEAM IN LANCASTER COUNTY

CHAPTER 1: K-1 MATH EXERCISES

CHAPTER 2: K-2 MATH EXERCISES

ADDITIONS

1.	22 + 11 = _____
2.	6 + 12 = _____
3.	49 + 32 = _____
4.	21 + 28 = _____
5.	5 + 31 = _____
6.	27 + 36 = _____
7.	16 + 44 = _____
8.	28 + 30 = _____
9.	16 + 33 = _____
10.	34 + 29 = _____
11.	40 + 8 = _____
12.	37 + 6 = _____
13.	28 + 37 = _____
14.	34 + 47 = _____

15.   42 + 8 = _____

16.   15 + 43 = _____

17.   33 + 13 = _____

18.   18 + 5 = _____

19.   27 + 32 = _____

20.   48 + 47 = _____

21.   8 + 50 = _____

22.   32 + 27 = _____

23.   47 + 32 = _____

24.   34 + 47 = _____

25.   20 + 46 = _____

26.   46 + 26 = _____

27.   12 + 24 = _____

28.   6 + 49 = _____

29.   34 + 23 = _____

30.   15 + 22 = _____

31.   13 + 35 = _____

32.   35 + 45 = _____

33. 46 + 44 = _____

34. 27 + 7 = _____

35. 21 + 42 = _____

36. 25 + 43 = _____

37. 46 + 12 = _____

38. 28 + 26 = _____

39. 32 + 15 = _____

40. 22 + 37 = _____

41. 17 + 45 = _____

42. 21 + 34 = _____

43. 34 + 46 = _____

44. 48 + 31 = _____

45. 24 + 41 = _____

46. 43 + 32 = _____

47. 22 + 15 = _____

48. 25 + 17 = _____

49. 31 + 6 = _____

50. 42 + 12 = _____

51.   16 + 40 = _____

52.   13 + 26 = _____

53.   50 + 41 = _____

54.   3 + 44 = _____

55.   2 + 32 = _____

56.   39 + 28 = _____

57.   36 + 33 = _____

58.   4 + 5 = _____

59.   17 + 25 = _____

60.   15 + 42 = _____

61.   38 + 33 = _____

62.   5 + 14 = _____

63.   39 + 38 = _____

64.   17 + 9 = _____

65.   50 + 46 = _____

66.   36 + 29 = _____

67.   48 + 12 = _____

68.   33 + 43 = _____

69.	37 + 39 = ____
70.	29 + 43 = ____
71.	10 + 37 = ____
72.	18 + 8 = ____
73.	4 + 39 = ____
74.	32 + 21 = ____
75.	24 + 28 = ____
76.	9 + 48 = ____
77.	47 + 49 = ____
78.	50 + 4 = ____
79.	14 + 49 = ____
80.	32 + 8 = ____
81.	39 + 37 = ____
82.	24 + 13 = ____
83.	37 + 40 = ____
84.	8 + 14 = ____
85.	15 + 47 = ____
86.	33 + 22 = ____

87.	19 + 25 = ____
88.	10 + 38 = ____
89.	39 + 6 = ____
90.	5 + 13 = ____
91.	17 + 39 = ____
92.	20 + 47 = ____
93.	41 + 38 = ____
94.	22 + 9 = ____
95.	39 + 46 = ____
96.	5 + 22 = ____
97.	43 + 6 = ____
98.	34 + 36 = ____
99.	50 + 49 = ____
100.	23 + 8 = ____
1.	22 + 11 = 33
2.	6 + 12 = 18
3.	49 + 32 = 81
4.	21 + 28 = 49

5.	$5 + 31 = 36$
6.	$27 + 36 = 63$
7.	$16 + 44 = 60$
8.	$28 + 30 = 58$
9.	$16 + 33 = 49$
10.	$34 + 29 = 63$
11.	$40 + 8 = 48$
12.	$37 + 6 = 43$
13.	$28 + 37 = 65$
14.	$34 + 47 = 81$
15.	$42 + 8 = 50$
16.	$15 + 43 = 58$
17.	$33 + 13 = 46$
18.	$18 + 5 = 23$
19.	$27 + 32 = 59$
20.	$48 + 47 = 95$
21.	$8 + 50 = 58$
22.	$32 + 27 = 59$

CHAPTER 2: K-2 MATH EXERCISES

23.	47 + 32 = 79
24.	34 + 47 = 81
25.	20 + 46 = 66
26.	46 + 26 = 72
27.	12 + 24 = 36
28.	6 + 49 = 55
29.	34 + 23 = 57
30.	15 + 22 = 37
31.	13 + 35 = 48
32.	35 + 45 = 80
33.	46 + 44 = 90
34.	27 + 7 = 34
35.	21 + 42 = 63
36.	25 + 43 = 68
37.	46 + 12 = 58
38.	28 + 26 = 54
39.	32 + 15 = 47
40.	22 + 37 = 59

41.	17 + 45 = 62
42.	21 + 34 = 55
43.	34 + 46 = 80
44.	48 + 31 = 79
45.	24 + 41 = 65
46.	43 + 32 = 75
47.	22 + 15 = 37
48.	25 + 17 = 42
49.	31 + 6 = 37
50.	42 + 12 = 54
51.	16 + 40 = 56
52.	13 + 26 = 39
53.	50 + 41 = 91
54.	3 + 44 = 47
55.	2 + 32 = 34
56.	39 + 28 = 67
57.	36 + 33 = 69
58.	4 + 5 = 9

59.	17 + 25 = 42
60.	15 + 42 = 57
61.	38 + 33 = 71
62.	5 + 14 = 19
63.	39 + 38 = 77
64.	17 + 9 = 26
65.	50 + 46 = 96
66.	36 + 29 = 65
67.	48 + 12 = 60
68.	33 + 43 = 76
69.	37 + 39 = 76
70.	29 + 43 = 72
71.	10 + 37 = 47
72.	18 + 8 = 26
73.	4 + 39 = 43
74.	32 + 21 = 53
75.	24 + 28 = 52
76.	9 + 48 = 57

77.	47 + 49 = 96
78.	50 + 4 = 54
79.	14 + 49 = 63
80.	32 + 8 = 40
81.	39 + 37 = 76
82.	24 + 13 = 37
83.	37 + 40 = 77
84.	8 + 14 = 22
85.	15 + 47 = 62
86.	33 + 22 = 55
87.	19 + 25 = 44
88.	10 + 38 = 48
89.	39 + 6 = 45
90.	5 + 13 = 18
91.	17 + 39 = 56
92.	20 + 47 = 67
93.	41 + 38 = 79
94.	22 + 9 = 31

95.	39 + 46 = 85
96.	5 + 22 = 27
97.	43 + 6 = 49
98.	34 + 36 = 70
99.	50 + 49 = 99
100.	23 + 8 = 31
1.	____ + 11 = 33
2.	____ + 12 = 18
3.	____ + 32 = 81
4.	____ + 28 = 49
5.	____ + 31 = 36
6.	____ + 36 = 63
7.	____ + 44 = 60
8.	____ + 30 = 58
9.	____ + 33 = 49
10.	____ + 29 = 63
11.	____ + 8 = 48
12.	____ + 6 = 43

13. _____ + 37 = 65

14. _____ + 47 = 81

15. _____ + 8 = 50

16. _____ + 43 = 58

17. _____ + 13 = 46

18. _____ + 5 = 23

19. _____ + 32 = 59

20. _____ + 47 = 95

21. _____ + 50 = 58

22. _____ + 27 = 59

23. _____ + 32 = 79

24. _____ + 47 = 81

25. _____ + 46 = 66

26. _____ + 26 = 72

27. _____ + 24 = 36

28. _____ + 49 = 55

29. _____ + 23 = 57

30. _____ + 22 = 37

31.   ____ + 35 = 48

32.   ____ + 45 = 80

33.   ____ + 44 = 90

34.   ____ + 7 = 34

35.   ____ + 42 = 63

36.   ____ + 43 = 68

37.   ____ + 12 = 58

38.   ____ + 26 = 54

39.   ____ + 15 = 47

40.   ____ + 37 = 59

41.   ____ + 45 = 62

42.   ____ + 34 = 55

43.   ____ + 46 = 80

44.   ____ + 31 = 79

45.   ____ + 41 = 65

46.   ____ + 32 = 75

47.   ____ + 15 = 37

48.   ____ + 17 = 42

49.	_____ + 6 = 37
50.	_____ + 12 = 54
51.	_____ + 40 = 56
52.	_____ + 26 = 39
53.	_____ + 41 = 91
54.	_____ + 44 = 47
55.	_____ + 32 = 34
56.	_____ + 28 = 67
57.	_____ + 33 = 69
58.	_____ + 5 = 9
59.	_____ + 25 = 42
60.	_____ + 42 = 57
61.	_____ + 33 = 71
62.	_____ + 14 = 19
63.	_____ + 38 = 77
64.	_____ + 9 = 26
65.	_____ + 46 = 96
66.	_____ + 29 = 65

67.	____ + 12 = 60
68.	____ + 43 = 76
69.	____ + 39 = 76
70.	____ + 43 = 72
71.	____ + 37 = 47
72.	____ + 8 = 26
73.	____ + 39 = 43
74.	____ + 21 = 53
75.	____ + 28 = 52
76.	____ + 48 = 57
77.	____ + 49 = 96
78.	____ + 4 = 54
79.	____ + 49 = 63
80.	____ + 8 = 40
81.	____ + 37 = 76
82.	____ + 13 = 37
83.	____ + 40 = 77
84.	____ + 14 = 22

85.	_____ + 47 = 62
86.	_____ + 22 = 55
87.	_____ + 25 = 44
88.	_____ + 38 = 48
89.	_____ + 6 = 45
90.	_____ + 13 = 18
91.	_____ + 39 = 56
92.	_____ + 47 = 67
93.	_____ + 38 = 79
94.	_____ + 9 = 31
95.	_____ + 46 = 85
96.	_____ + 22 = 27
97.	_____ + 6 = 49
98.	_____ + 36 = 70
99.	_____ + 49 = 99
100.	_____ + 8 = 31
1.	22 + _____ = 33
2.	6 + _____ = 18

3.	$49 + \underline{\hspace{1cm}} = 81$
4.	$21 + \underline{\hspace{1cm}} = 49$
5.	$5 + \underline{\hspace{1cm}} = 36$
6.	$27 + \underline{\hspace{1cm}} = 63$
7.	$16 + \underline{\hspace{1cm}} = 60$
8.	$28 + \underline{\hspace{1cm}} = 58$
9.	$16 + \underline{\hspace{1cm}} = 49$
10.	$34 + \underline{\hspace{1cm}} = 63$
11.	$40 + \underline{\hspace{1cm}} = 48$
12.	$37 + \underline{\hspace{1cm}} = 43$
13.	$28 + \underline{\hspace{1cm}} = 65$
14.	$34 + \underline{\hspace{1cm}} = 81$
15.	$42 + \underline{\hspace{1cm}} = 50$
16.	$15 + \underline{\hspace{1cm}} = 58$
17.	$33 + \underline{\hspace{1cm}} = 46$
18.	$18 + \underline{\hspace{1cm}} = 23$
19.	$27 + \underline{\hspace{1cm}} = 59$
20.	$48 + \underline{\hspace{1cm}} = 95$

21.   8 + ____ = 58

22.   32 + ____ = 59

23.   47 + ____ = 79

24.   34 + ____ = 81

25.   20 + ____ = 66

26.   46 + ____ = 72

27.   12 + ____ = 36

28.   6 + ____ = 55

29.   34 + ____ = 57

30.   15 + ____ = 37

31.   13 + ____ = 48

32.   35 + ____ = 80

33.   46 + ____ = 90

34.   27 + ____ = 34

35.   21 + ____ = 63

36.   25 + ____ = 68

37.   46 + ____ = 58

38.   28 + ____ = 54

39.	32 + _____ = 47
40.	22 + _____ = 59
41.	17 + _____ = 62
42.	21 + _____ = 55
43.	34 + _____ = 80
44.	48 + _____ = 79
45.	24 + _____ = 65
46.	43 + _____ = 75
47.	22 + _____ = 37
48.	25 + _____ = 42
49.	31 + _____ = 37
50.	42 + _____ = 54
51.	16 + _____ = 56
52.	13 + _____ = 39
53.	50 + _____ = 91
54.	3 + _____ = 47
55.	2 + _____ = 34
56.	39 + _____ = 67

57. $36 + \rule{1cm}{0.15mm} = 69$

58. $4 + \rule{1cm}{0.15mm} = 9$

59. $17 + \rule{1cm}{0.15mm} = 42$

60. $15 + \rule{1cm}{0.15mm} = 57$

61. $38 + \rule{1cm}{0.15mm} = 71$

62. $5 + \rule{1cm}{0.15mm} = 19$

63. $39 + \rule{1cm}{0.15mm} = 77$

64. $17 + \rule{1cm}{0.15mm} = 26$

65. $50 + \rule{1cm}{0.15mm} = 96$

66. $36 + \rule{1cm}{0.15mm} = 65$

67. $48 + \rule{1cm}{0.15mm} = 60$

68. $33 + \rule{1cm}{0.15mm} = 76$

69. $37 + \rule{1cm}{0.15mm} = 76$

70. $29 + \rule{1cm}{0.15mm} = 72$

71. $10 + \rule{1cm}{0.15mm} = 47$

72. $18 + \rule{1cm}{0.15mm} = 26$

73. $4 + \rule{1cm}{0.15mm} = 43$

74. $32 + \rule{1cm}{0.15mm} = 53$

75.   24 + _____ = 52

76.   9 + _____ = 57

77.   47 + _____ = 96

78.   50 + _____ = 54

79.   14 + _____ = 63

80.   32 + _____ = 40

81.   39 + _____ = 76

82.   24 + _____ = 37

83.   37 + _____ = 77

84.   8 + _____ = 22

85.   15 + _____ = 62

86.   33 + _____ = 55

87.   19 + _____ = 44

88.   10 + _____ = 48

89.   39 + _____ = 45

90.   5 + _____ = 18

91.   17 + _____ = 56

92.   20 + _____ = 67

93.	$41 + \rule{1.5em}{0.4pt} = 79$
94.	$22 + \rule{1.5em}{0.4pt} = 31$
95.	$39 + \rule{1.5em}{0.4pt} = 85$
96.	$5 + \rule{1.5em}{0.4pt} = 27$
97.	$43 + \rule{1.5em}{0.4pt} = 49$
98.	$34 + \rule{1.5em}{0.4pt} = 70$
99.	$50 + \rule{1.5em}{0.4pt} = 99$
100.	$23 + \rule{1.5em}{0.4pt} = 31$
1.	Twenty-Two + Eleven = \rule{2em}{0.4pt}
2.	Six + Twelve = \rule{2em}{0.4pt}
3.	Forty-Nine + Thirty-Two = \rule{2em}{0.4pt}
4.	Twenty-One + Twenty-Eight = \rule{2em}{0.4pt}
5.	Five + Thirty-One = \rule{2em}{0.4pt}
6.	Twenty-Seven + Thirty-Six = \rule{2em}{0.4pt}
7.	Sixteen + Forty-Four = \rule{2em}{0.4pt}
8.	Twenty-Eight + Thirty = \rule{2em}{0.4pt}
9.	Sixteen + Thirty-Three = \rule{2em}{0.4pt}
10.	Thirty-Four + Twenty-Nine = \rule{2em}{0.4pt}

1.	Twenty-Two + Eleven = Thirty-Three
2.	Six + Twelve = Eighteen
3.	Forty-Nine + Thirty-Two = Eighty-One
4.	Twenty-One + Twenty-Eight = Forty-Nine
5.	Five + Thirty-One = Thirty-Six
6.	Twenty-Seven + Thirty-Six = Sixty-Three
7.	Sixteen + Forty-Four = Sixty
8.	Twenty-Eight + Thirty = Fifty-Eight
9.	Sixteen + Thirty-Three = Forty-Nine
10.	Thirty-Four + Twenty-Nine = Sixty-Three
1.	_____ + Eleven = Thirty-Three
2.	_____ + Twelve = Eighteen
3.	_____ + Thirty-Two = Eighty-One
4.	_____ + Twenty-Eight = Forty-Nine
5.	_____ + Thirty-One = Thirty-Six
6.	_____ + Thirty-Six = Sixty-Three

7. ____ + Forty-Four = Sixty

8. ____ + Thirty = Fifty-Eight

9. ____ + Thirty-Three = Forty-Nine

10. ____ + Twenty-Nine = Sixty-Three

1. Twenty-Two + ____ = Thirty-Three

2. Six + ____ = Eighteen

3. Forty-Nine + ____ = Eighty-One

4. Twenty-One + ____ = Forty-Nine

5. Five + ____ = Thirty-Six

6. Twenty-Seven + ____ = Sixty-Three

7. Sixteen + ____ = Sixty

8. Twenty-Eight + ____ = Fifty-Eight

9. Sixteen + ____ = Forty-Nine

10. Thirty-Four + ____ = Sixty-Three

FOUNTAIN ON 6TH AVENUE IN NEW YORK CITY, NEW YORK

CHAPTER 2: K-2 MATH EXERCISES

SUBTRACTIONS

1.	48 - 35 = _____
2.	41 - 12 = _____
3.	49 - 43 = _____
4.	50 - 28 = _____
5.	36 - 20 = _____
6.	46 - 34 = _____
7.	45 - 10 = _____
8.	43 - 6 = _____
9.	49 - 43 = _____
10.	43 - 36 = _____
11.	24 - 12 = _____
12.	46 - 24 = _____
13.	49 - 43 = _____
14.	42 - 18 = _____
15.	40 - 4 = _____
16.	49 - 39 = _____

17.   7 - 4 = ____

18.   36 - 26 = ____

19.   35 - 15 = ____

20.   42 - 26 = ____

21.   32 - 12 = ____

22.   39 - 31 = ____

23.   30 - 6 = ____

24.   18 - 3 = ____

25.   46 - 38 = ____

26.   22 - 17 = ____

27.   42 - 39 = ____

28.   50 - 42 = ____

29.   40 - 2 = ____

30.   47 - 8 = ____

31.   48 - 10 = ____

32.   19 - 3 = ____

33.   29 - 15 = ____

34.   45 - 25 = ____

35.	23 - 22 = ____
36.	27 - 14 = ____
37.	30 - 22 = ____
38.	29 - 23 = ____
39.	35 - 13 = ____
40.	14 - 4 = ____
41.	46 - 19 = ____
42.	40 - 11 = ____
43.	34 - 30 = ____
44.	49 - 45 = ____
45.	35 - 28 = ____
46.	31 - 10 = ____
47.	39 - 9 = ____
48.	30 - 28 = ____
49.	31 - 25 = ____
50.	47 - 39 = ____
51.	31 - 28 = ____
52.	48 - 19 = ____

53.	20 - 13 = _____
54.	26 - 21 = _____
55.	37 - 11 = _____
56.	34 - 7 = _____
57.	45 - 18 = _____
58.	32 - 31 = _____
59.	31 - 30 = _____
60.	45 - 13 = _____
61.	37 - 18 = _____
62.	49 - 33 = _____
63.	17 - 12 = _____
64.	14 - 5 = _____
65.	48 - 45 = _____
66.	47 - 29 = _____
67.	14 - 12 = _____
68.	50 - 42 = _____
69.	47 - 14 = _____
70.	50 - 3 = _____

71.   13 - 2 = _____

72.   25 - 20 = _____

73.   24 - 19 = _____

74.   38 - 34 = _____

75.   45 - 18 = _____

76.   10 - 6 = _____

77.   27 - 8 = _____

78.   28 - 7 = _____

79.   31 - 24 = _____

80.   23 - 13 = _____

81.   18 - 11 = _____

82.   20 - 2 = _____

83.   35 - 6 = _____

84.   33 - 32 = _____

85.   48 - 2 = _____

86.   43 - 37 = _____

87.   41 - 2 = _____

88.   27 - 8 = _____

89.	45 - 12 = ____
90.	39 - 17 = ____
91.	44 - 37 = ____
92.	7 - 2 = ____
93.	44 - 29 = ____
94.	30 - 7 = ____
95.	31 - 2 = ____
96.	34 - 33 = ____
97.	32 - 24 = ____
98.	25 - 3 = ____
99.	48 - 6 = ____
100.	37 - 3 = ____
1.	48 - 35 = 13
2.	41 - 12 = 29
3.	49 - 43 = 6
4.	50 - 28 = 22
5.	36 - 20 = 16
6.	46 - 34 = 12

7.	45 - 10 = 35
8.	43 - 6 = 37
9.	49 - 43 = 6
10.	43 - 36 = 7
11.	24 - 12 = 12
12.	46 - 24 = 22
13.	49 - 43 = 6
14.	42 - 18 = 24
15.	40 - 4 = 36
16.	49 - 39 = 10
17.	7 - 4 = 3
18.	36 - 26 = 10
19.	35 - 15 = 20
20.	42 - 26 = 16
21.	32 - 12 = 20
22.	39 - 31 = 8
23.	30 - 6 = 24
24.	18 - 3 = 15

25.	46 - 38 = 8
26.	22 - 17 = 5
27.	42 - 39 = 3
28.	50 - 42 = 8
29.	40 - 2 = 38
30.	47 - 8 = 39
31.	48 - 10 = 38
32.	19 - 3 = 16
33.	29 - 15 = 14
34.	45 - 25 = 20
35.	23 - 22 = 1
36.	27 - 14 = 13
37.	30 - 22 = 8
38.	29 - 23 = 6
39.	35 - 13 = 22
40.	14 - 4 = 10
41.	46 - 19 = 27
42.	40 - 11 = 29

43.	34 - 30 = 4
44.	49 - 45 = 4
45.	35 - 28 = 7
46.	31 - 10 = 21
47.	39 - 9 = 30
48.	30 - 28 = 2
49.	31 - 25 = 6
50.	47 - 39 = 8
51.	31 - 28 = 3
52.	48 - 19 = 29
53.	20 - 13 = 7
54.	26 - 21 = 5
55.	37 - 11 = 26
56.	34 - 7 = 27
57.	45 - 18 = 27
58.	32 - 31 = 1
59.	31 - 30 = 1
60.	45 - 13 = 32

61.	37 - 18 = 19
62.	49 - 33 = 16
63.	17 - 12 = 5
64.	14 - 5 = 9
65.	48 - 45 = 3
66.	47 - 29 = 18
67.	14 - 12 = 2
68.	50 - 42 = 8
69.	47 - 14 = 33
70.	50 - 3 = 47
71.	13 - 2 = 11
72.	25 - 20 = 5
73.	24 - 19 = 5
74.	38 - 34 = 4
75.	45 - 18 = 27
76.	10 - 6 = 4
77.	27 - 8 = 19
78.	28 - 7 = 21

79.	31 - 24 = 7
80.	23 - 13 = 10
81.	18 - 11 = 7
82.	20 - 2 = 18
83.	35 - 6 = 29
84.	33 - 32 = 1
85.	48 - 2 = 46
86.	43 - 37 = 6
87.	41 - 2 = 39
88.	27 - 8 = 19
89.	45 - 12 = 33
90.	39 - 17 = 22
91.	44 - 37 = 7
92.	7 - 2 = 5
93.	44 - 29 = 15
94.	30 - 7 = 23
95.	31 - 2 = 29
96.	34 - 33 = 1

97.	32 - 24 = 8
98.	25 - 3 = 22
99.	48 - 6 = 42
100.	37 - 3 = 34
1.	____ - 35 = 13
2.	____ - 12 = 29
3.	____ - 43 = 6
4.	____ - 28 = 22
5.	____ - 20 = 16
6.	____ - 34 = 12
7.	____ - 10 = 35
8.	____ - 6 = 37
9.	____ - 43 = 6
10.	____ - 36 = 7
11.	____ - 12 = 12
12.	____ - 24 = 22
13.	____ - 43 = 6
14.	____ - 18 = 24

15. _____ - 4 = 36

16. _____ - 39 = 10

17. _____ - 4 = 3

18. _____ - 26 = 10

19. _____ - 15 = 20

20. _____ - 26 = 16

21. _____ - 12 = 20

22. _____ - 31 = 8

23. _____ - 6 = 24

24. _____ - 3 = 15

25. _____ - 38 = 8

26. _____ - 17 = 5

27. _____ - 39 = 3

28. _____ - 42 = 8

29. _____ - 2 = 38

30. _____ - 8 = 39

31. _____ - 10 = 38

32. _____ - 3 = 16

33.   _____ - 15 = 14

34.   _____ - 25 = 20

35.   _____ - 22 = 1

36.   _____ - 14 = 13

37.   _____ - 22 = 8

38.   _____ - 23 = 6

39.   _____ - 13 = 22

40.   _____ - 4 = 10

41.   _____ - 19 = 27

42.   _____ - 11 = 29

43.   _____ - 30 = 4

44.   _____ - 45 = 4

45.   _____ - 28 = 7

46.   _____ - 10 = 21

47.   _____ - 9 = 30

48.   _____ - 28 = 2

49.   _____ - 25 = 6

50.   _____ - 39 = 8

51.   _____ - 28 = 3

52.   _____ - 19 = 29

53.   _____ - 13 = 7

54.   _____ - 21 = 5

55.   _____ - 11 = 26

56.   _____ - 7 = 27

57.   _____ - 18 = 27

58.   _____ - 31 = 1

59.   _____ - 30 = 1

60.   _____ - 13 = 32

61.   _____ - 18 = 19

62.   _____ - 33 = 16

63.   _____ - 12 = 5

64.   _____ - 5 = 9

65.   _____ - 45 = 3

66.   _____ - 29 = 18

67.   _____ - 12 = 2

68.   _____ - 42 = 8

69.	____ - 14 = 33
70.	____ - 3 = 47
71.	____ - 2 = 11
72.	____ - 20 = 5
73.	____ - 19 = 5
74.	____ - 34 = 4
75.	____ - 18 = 27
76.	____ - 6 = 4
77.	____ - 8 = 19
78.	____ - 7 = 21
79.	____ - 24 = 7
80.	____ - 13 = 10
81.	____ - 11 = 7
82.	____ - 2 = 18
83.	____ - 6 = 29
84.	____ - 32 = 1
85.	____ - 2 = 46
86.	____ - 37 = 6

87. _____ - 2 = 39

88. _____ - 8 = 19

89. _____ - 12 = 33

90. _____ - 17 = 22

91. _____ - 37 = 7

92. _____ - 2 = 5

93. _____ - 29 = 15

94. _____ - 7 = 23

95. _____ - 2 = 29

96. _____ - 33 = 1

97. _____ - 24 = 8

98. _____ - 3 = 22

99. _____ - 6 = 42

100. _____ - 3 = 34

1. 48 - _____ = 13

2. 41 - _____ = 29

3. 49 - _____ = 6

4. 50 - _____ = 22

5.   36 - ____ = 16

6.   46 - ____ = 12

7.   45 - ____ = 35

8.   43 - ____ = 37

9.   49 - ____ = 6

10.   43 - ____ = 7

11.   24 - ____ = 12

12.   46 - ____ = 22

13.   49 - ____ = 6

14.   42 - ____ = 24

15.   40 - ____ = 36

16.   49 - ____ = 10

17.   7 - ____ = 3

18.   36 - ____ = 10

19.   35 - ____ = 20

20.   42 - ____ = 16

21.   32 - ____ = 20

22.   39 - ____ = 8

23.   30 - ____ = 24

24.   18 - ____ = 15

25.   46 - ____ = 8

26.   22 - ____ = 5

27.   42 - ____ = 3

28.   50 - ____ = 8

29.   40 - ____ = 38

30.   47 - ____ = 39

31.   48 - ____ = 38

32.   19 - ____ = 16

33.   29 - ____ = 14

34.   45 - ____ = 20

35.   23 - ____ = 1

36.   27 - ____ = 13

37.   30 - ____ = 8

38.   29 - ____ = 6

39.   35 - ____ = 22

40.   14 - ____ = 10

CHAPTER 2: K-2 MATH EXERCISES

41.   46 - _____ = 27

42.   40 - _____ = 29

43.   34 - _____ = 4

44.   49 - _____ = 4

45.   35 - _____ = 7

46.   31 - _____ = 21

47.   39 - _____ = 30

48.   30 - _____ = 2

49.   31 - _____ = 6

50.   47 - _____ = 8

51.   31 - _____ = 3

52.   48 - _____ = 29

53.   20 - _____ = 7

54.   26 - _____ = 5

55.   37 - _____ = 26

56.   34 - _____ = 27

57.   45 - _____ = 27

58.   32 - _____ = 1

59.   31 - _____ = 1

60.   45 - _____ = 32

61.   37 - _____ = 19

62.   49 - _____ = 16

63.   17 - _____ = 5

64.   14 - _____ = 9

65.   48 - _____ = 3

66.   47 - _____ = 18

67.   14 - _____ = 2

68.   50 - _____ = 8

69.   47 - _____ = 33

70.   50 - _____ = 47

71.   13 - _____ = 11

72.   25 - _____ = 5

73.   24 - _____ = 5

74.   38 - _____ = 4

75.   45 - _____ = 27

76.   10 - _____ = 4

77. $27 - \underline{\hspace{1cm}} = 19$

78. $28 - \underline{\hspace{1cm}} = 21$

79. $31 - \underline{\hspace{1cm}} = 7$

80. $23 - \underline{\hspace{1cm}} = 10$

81. $18 - \underline{\hspace{1cm}} = 7$

82. $20 - \underline{\hspace{1cm}} = 18$

83. $35 - \underline{\hspace{1cm}} = 29$

84. $33 - \underline{\hspace{1cm}} = 1$

85. $48 - \underline{\hspace{1cm}} = 46$

86. $43 - \underline{\hspace{1cm}} = 6$

87. $41 - \underline{\hspace{1cm}} = 39$

88. $27 - \underline{\hspace{1cm}} = 19$

89. $45 - \underline{\hspace{1cm}} = 33$

90. $39 - \underline{\hspace{1cm}} = 22$

91. $44 - \underline{\hspace{1cm}} = 7$

92. $7 - \underline{\hspace{1cm}} = 5$

93. $44 - \underline{\hspace{1cm}} = 15$

94. $30 - \underline{\hspace{1cm}} = 23$

95.	31 - ____ = 29
96.	34 - ____ = 1
97.	32 - ____ = 8
98.	25 - ____ = 22
99.	48 - ____ = 42
100.	37 - ____ = 34
1.	Forty-Eight - Thirty-Five = ____
2.	Forty-One - Twelve = ____
3.	Forty-Nine - Forty-Three = ____
4.	Fifty - Twenty-Eight = ____
5.	Thirty-Six - Twenty = ____
6.	Forty-Six - Thirty-Four = ____
7.	Forty-Five - Ten = ____
8.	Forty-Three - Six = ____
9.	Forty-Nine - Forty-Three = ____
10.	Forty-Three - Thirty-Six = ____
1.	Forty-Eight - Thirty-Five = Thirteen
2.	Forty-One - Twelve = Twenty-Nine

3.	Forty-Nine - Forty-Three = Six
4.	Fifty - Twenty-Eight = Twenty-Two
5.	Thirty-Six - Twenty = Sixteen
6.	Forty-Six - Thirty-Four = Twelve
7.	Forty-Five - Ten = Thirty-Five
8.	Forty-Three - Six = Thirty-Seven
9.	Forty-Nine - Forty-Three = Six
10.	Forty-Three - Thirty-Six = Seven
1.	_____ - Thirty-Five = Thirteen
2.	_____ - Twelve = Twenty-Nine
3.	_____ - Forty-Three = Six
4.	_____ - Twenty-Eight = Twenty-Two
5.	_____ - Twenty = Sixteen
6.	_____ - Thirty-Four = Twelve
7.	_____ - Ten = Thirty-Five
8.	_____ - Six = Thirty-Seven
9.	_____ - Forty-Three = Six
10.	_____ - Thirty-Six = Seven

1. Forty-Eight - _____ = Thirteen

2. Forty-One - _____ = Twenty-Nine

3. Forty-Nine - _____ = Six

4. Fifty - _____ = Twenty-Two

5. Thirty-Six - _____ = Sixteen

6. Forty-Six - _____ = Twelve

7. Forty-Five - _____ = Thirty-Five

8. Forty-Three - _____ = Thirty-Seven

9. Forty-Nine - _____ = Six

10. Forty-Three - _____ = Seven

RED CUBE SCULPTURE ON BROADWAY NEAR WORLD TRADE CENTER IN NYC

CHAPTER 2: K-2 MATH EXERCISES

MULTIPLICATIONS

1.	10 x 10 = _____
2.	7 x 6 = _____
3.	4 x 3 = _____
4.	3 x 6 = _____
5.	10 x 2 = _____
6.	2 x 6 = _____
7.	6 x 4 = _____
8.	3 x 3 = _____
9.	4 x 8 = _____
10.	6 x 3 = _____
11.	5 x 6 = _____
12.	10 x 8 = _____
13.	9 x 10 = _____
14.	8 x 9 = _____
15.	7 x 6 = _____
16.	6 x 7 = _____

17.   3 x 2 = _____

18.   6 x 3 = _____

19.   5 x 7 = _____

20.   2 x 3 = _____

21.   6 x 7 = _____

22.   5 x 3 = _____

23.   2 x 5 = _____

24.   7 x 10 = _____

25.   5 x 6 = _____

26.   6 x 9 = _____

27.   5 x 4 = _____

28.   2 x 6 = _____

29.   3 x 7 = _____

30.   7 x 3 = _____

31.   10 x 10 = _____

32.   8 x 7 = _____

33.   4 x 10 = _____

34.   9 x 9 = _____

35.	6 x 7 = ____
36.	8 x 9 = ____
37.	9 x 7 = ____
38.	7 x 10 = ____
39.	6 x 7 = ____
40.	4 x 8 = ____
41.	7 x 6 = ____
42.	9 x 7 = ____
43.	10 x 8 = ____
44.	4 x 9 = ____
45.	9 x 7 = ____
46.	4 x 3 = ____
47.	2 x 6 = ____
48.	5 x 4 = ____
49.	2 x 6 = ____
50.	7 x 4 = ____
1.	10 x 10 = 100
2.	7 x 6 = 42

3.	4 x 3 = 12
4.	3 x 6 = 18
5.	10 x 2 = 20
6.	2 x 6 = 12
7.	6 x 4 = 24
8.	3 x 3 = 9
9.	4 x 8 = 32
10.	6 x 3 = 18
11.	5 x 6 = 30
12.	10 x 8 = 80
13.	9 x 10 = 90
14.	8 x 9 = 72
15.	7 x 6 = 42
16.	6 x 7 = 42
17.	3 x 2 = 6
18.	6 x 3 = 18
19.	5 x 7 = 35
20.	2 x 3 = 6

21.	6 x 7 = 42
22.	5 x 3 = 15
23.	2 x 5 = 10
24.	7 x 10 = 70
25.	5 x 6 = 30
26.	6 x 9 = 54
27.	5 x 4 = 20
28.	2 x 6 = 12
29.	3 x 7 = 21
30.	7 x 3 = 21
31.	10 x 10 = 100
32.	8 x 7 = 56
33.	4 x 10 = 40
34.	9 x 9 = 81
35.	6 x 7 = 42
36.	8 x 9 = 72
37.	9 x 7 = 63
38.	7 x 10 = 70

39.	6 x 7 = 42
40.	4 x 8 = 32
41.	7 x 6 = 42
42.	9 x 7 = 63
43.	10 x 8 = 80
44.	4 x 9 = 36
45.	9 x 7 = 63
46.	4 x 3 = 12
47.	2 x 6 = 12
48.	5 x 4 = 20
49.	2 x 6 = 12
50.	7 x 4 = 28
1.	_____ x 10 = 100
2.	_____ x 6 = 42
3.	_____ x 3 = 12
4.	_____ x 6 = 18
5.	_____ x 2 = 20
6.	_____ x 6 = 12

7. _____ x 4 = 24

8. _____ x 3 = 9

9. _____ x 8 = 32

10. _____ x 3 = 18

11. _____ x 6 = 30

12. _____ x 8 = 80

13. _____ x 10 = 90

14. _____ x 9 = 72

15. _____ x 6 = 42

16. _____ x 7 = 42

17. _____ x 2 = 6

18. _____ x 3 = 18

19. _____ x 7 = 35

20. _____ x 3 = 6

21. _____ x 7 = 42

22. _____ x 3 = 15

23. _____ x 5 = 10

24. _____ x 10 = 70

25.   _____ x 6 = 30

26.   _____ x 9 = 54

27.   _____ x 4 = 20

28.   _____ x 6 = 12

29.   _____ x 7 = 21

30.   _____ x 3 = 21

31.   _____ x 10 = 100

32.   _____ x 7 = 56

33.   _____ x 10 = 40

34.   _____ x 9 = 81

35.   _____ x 7 = 42

36.   _____ x 9 = 72

37.   _____ x 7 = 63

38.   _____ x 10 = 70

39.   _____ x 7 = 42

40.   _____ x 8 = 32

41.   _____ x 6 = 42

42.   _____ x 7 = 63

43. _____ x 8 = 80

44. _____ x 9 = 36

45. _____ x 7 = 63

46. _____ x 3 = 12

47. _____ x 6 = 12

48. _____ x 4 = 20

49. _____ x 6 = 12

50. _____ x 4 = 28

1. 10 x _____ = 100

2. 7 x _____ = 42

3. 4 x _____ = 12

4. 3 x _____ = 18

5. 10 x _____ = 20

6. 2 x _____ = 12

7. 6 x _____ = 24

8. 3 x _____ = 9

9. 4 x _____ = 32

10. 6 x _____ = 18

11.   5 x _____ = 30

12.   10 x _____ = 80

13.   9 x _____ = 90

14.   8 x _____ = 72

15.   7 x _____ = 42

16.   6 x _____ = 42

17.   3 x _____ = 6

18.   6 x _____ = 18

19.   5 x _____ = 35

20.   2 x _____ = 6

21.   6 x _____ = 42

22.   5 x _____ = 15

23.   2 x _____ = 10

24.   7 x _____ = 70

25.   5 x _____ = 30

26.   6 x _____ = 54

27.   5 x _____ = 20

28.   2 x _____ = 12

29.	3 x _____ = 21
30.	7 x _____ = 21
31.	10 x _____ = 100
32.	8 x _____ = 56
33.	4 x _____ = 40
34.	9 x _____ = 81
35.	6 x _____ = 42
36.	8 x _____ = 72
37.	9 x _____ = 63
38.	7 x _____ = 70
39.	6 x _____ = 42
40.	4 x _____ = 32
41.	7 x _____ = 42
42.	9 x _____ = 63
43.	10 x _____ = 80
44.	4 x _____ = 36
45.	9 x _____ = 63
46.	4 x _____ = 12

47.	2 x ____ = 12
48.	5 x ____ = 20
49.	2 x ____ = 12
50.	7 x ____ = 28
1.	Ten x Ten = ____
2.	Seven x Six = ____
3.	Four x Three = ____
4.	Three x Six = ____
5.	Ten x Two = ____
6.	Two x Six = ____
7.	Six x Four = ____
8.	Three x Three = ____
9.	Four x Eight = ____
10.	Six x Three = ____
1.	Ten x Ten = One Hundred
2.	Seven x Six = Forty-Two
3.	Four x Three = Twelve
4.	Three x Six = Eighteen

5.	Ten x Two = Twenty
6.	Two x Six = Twelve
7.	Six x Four = Twenty-Four
8.	Three x Three = Nine
9.	Four x Eight = Thirty-Two
10.	Six x Three = Eighteen
1.	_____ x Ten = One Hundred
2.	_____ x Six = Forty-Two
3.	_____ x Three = Twelve
4.	_____ x Six = Eighteen
5.	_____ x Two = Twenty
6.	_____ x Six = Twelve
7.	_____ x Four = Twenty-Four
8.	_____ x Three = Nine
9.	_____ x Eight = Thirty-Two
10.	_____ x Three = Eighteen
1.	Ten x _____ = One Hundred
2.	Seven x _____ = Forty-Two

3.	Four x _____ = Twelve
4.	Three x _____ = Eighteen
5.	Ten x _____ = Twenty
6.	Two x _____ = Twelve
7.	Six x _____ = Twenty-Four
8.	Three x _____ = Nine
9.	Four x _____ = Thirty-Two
10.	Six x _____ = Eighteen

FREEDOM TOWER (WORLD TRADE CENTER ONE) NEAR COMPLETION IN NYC

CHAPTER 2: K-2 MATH EXERCISES

DIVISIONS

1.	$10 \div 2 = $ _____
2.	$6 \div 3 = $ _____
3.	$12 \div 4 = $ _____
4.	$10 \div 5 = $ _____
5.	$6 \div 2 = $ _____
6.	$20 \div 4 = $ _____
7.	$8 \div 2 = $ _____
8.	$12 \div 4 = $ _____
9.	$14 \div 7 = $ _____
10.	$20 \div 5 = $ _____
11.	$14 \div 7 = $ _____
12.	$16 \div 4 = $ _____
13.	$12 \div 3 = $ _____
14.	$20 \div 2 = $ _____
15.	$14 \div 7 = $ _____
16.	$6 \div 3 = $ _____

17. $4 \div 2 =$ _____

18. $18 \div 9 =$ _____

19. $14 \div 2 =$ _____

20. $6 \div 3 =$ _____

21. $8 \div 2 =$ _____

22. $10 \div 5 =$ _____

23. $20 \div 4 =$ _____

24. $15 \div 3 =$ _____

25. $16 \div 2 =$ _____

26. $9 \div 3 =$ _____

27. $18 \div 9 =$ _____

28. $8 \div 4 =$ _____

29. $18 \div 9 =$ _____

30. $20 \div 4 =$ _____

31. $9 \div 3 =$ _____

32. $20 \div 2 =$ _____

33. $8 \div 4 =$ _____

34. $10 \div 5 =$ _____

35.	$16 \div 4 = $ _____
36.	$18 \div 3 = $ _____
37.	$16 \div 2 = $ _____
38.	$12 \div 4 = $ _____
39.	$10 \div 2 = $ _____
40.	$16 \div 4 = $ _____
41.	$15 \div 5 = $ _____
42.	$16 \div 8 = $ _____
43.	$9 \div 3 = $ _____
44.	$20 \div 2 = $ _____
45.	$14 \div 7 = $ _____
46.	$12 \div 2 = $ _____
47.	$18 \div 9 = $ _____
48.	$8 \div 4 = $ _____
49.	$16 \div 8 = $ _____
50.	$18 \div 6 = $ _____
1.	$10 \div 2 = 5$
2.	$6 \div 3 = 2$

3.	$12 \div 4 = 3$
4.	$10 \div 5 = 2$
5.	$6 \div 2 = 3$
6.	$20 \div 4 = 5$
7.	$8 \div 2 = 4$
8.	$12 \div 4 = 3$
9.	$14 \div 7 = 2$
10.	$20 \div 5 = 4$
11.	$14 \div 7 = 2$
12.	$16 \div 4 = 4$
13.	$12 \div 3 = 4$
14.	$20 \div 2 = 10$
15.	$14 \div 7 = 2$
16.	$6 \div 3 = 2$
17.	$4 \div 2 = 2$
18.	$18 \div 9 = 2$
19.	$14 \div 2 = 7$
20.	$6 \div 3 = 2$

21.	$8 \div 2 = 4$
22.	$10 \div 5 = 2$
23.	$20 \div 4 = 5$
24.	$15 \div 3 = 5$
25.	$16 \div 2 = 8$
26.	$9 \div 3 = 3$
27.	$18 \div 9 = 2$
28.	$8 \div 4 = 2$
29.	$18 \div 9 = 2$
30.	$20 \div 4 = 5$
31.	$9 \div 3 = 3$
32.	$20 \div 2 = 10$
33.	$8 \div 4 = 2$
34.	$10 \div 5 = 2$
35.	$16 \div 4 = 4$
36.	$18 \div 3 = 6$
37.	$16 \div 2 = 8$
38.	$12 \div 4 = 3$

39.	$10 \div 2 = 5$
40.	$16 \div 4 = 4$
41.	$15 \div 5 = 3$
42.	$16 \div 8 = 2$
43.	$9 \div 3 = 3$
44.	$20 \div 2 = 10$
45.	$14 \div 7 = 2$
46.	$12 \div 2 = 6$
47.	$18 \div 9 = 2$
48.	$8 \div 4 = 2$
49.	$16 \div 8 = 2$
50.	$18 \div 6 = 3$
1.	_____ $\div 2 = 5$
2.	_____ $\div 3 = 2$
3.	_____ $\div 4 = 3$
4.	_____ $\div 5 = 2$
5.	_____ $\div 2 = 3$
6.	_____ $\div 4 = 5$

7.   _____ $\div 2 = 4$

8.   _____ $\div 4 = 3$

9.   _____ $\div 7 = 2$

10.   _____ $\div 5 = 4$

11.   _____ $\div 7 = 2$

12.   _____ $\div 4 = 4$

13.   _____ $\div 3 = 4$

14.   _____ $\div 2 = 10$

15.   _____ $\div 7 = 2$

16.   _____ $\div 3 = 2$

17.   _____ $\div 2 = 2$

18.   _____ $\div 9 = 2$

19.   _____ $\div 2 = 7$

20.   _____ $\div 3 = 2$

21.   _____ $\div 2 = 4$

22.   _____ $\div 5 = 2$

23.   _____ $\div 4 = 5$

24.   _____ $\div 3 = 5$

25. _____ ÷ 2 = 8

26. _____ ÷ 3 = 3

27. _____ ÷ 9 = 2

28. _____ ÷ 4 = 2

29. _____ ÷ 9 = 2

30. _____ ÷ 4 = 5

31. _____ ÷ 3 = 3

32. _____ ÷ 2 = 10

33. _____ ÷ 4 = 2

34. _____ ÷ 5 = 2

35. _____ ÷ 4 = 4

36. _____ ÷ 3 = 6

37. _____ ÷ 2 = 8

38. _____ ÷ 4 = 3

39. _____ ÷ 2 = 5

40. _____ ÷ 4 = 4

41. _____ ÷ 5 = 3

42. _____ ÷ 8 = 2

43.	_____ ÷ 3 = 3
44.	_____ ÷ 2 = 10
45.	_____ ÷ 7 = 2
46.	_____ ÷ 2 = 6
47.	_____ ÷ 9 = 2
48.	_____ ÷ 4 = 2
49.	_____ ÷ 8 = 2
50.	_____ ÷ 6 = 3
1.	10 ÷ _____ = 5
2.	6 ÷ _____ = 2
3.	12 ÷ _____ = 3
4.	10 ÷ _____ = 2
5.	6 ÷ _____ = 3
6.	20 ÷ _____ = 5
7.	8 ÷ _____ = 4
8.	12 ÷ _____ = 3
9.	14 ÷ _____ = 2
10.	20 ÷ _____ = 4

11. $14 \div \underline{\hspace{1cm}} = 2$

12. $16 \div \underline{\hspace{1cm}} = 4$

13. $12 \div \underline{\hspace{1cm}} = 4$

14. $20 \div \underline{\hspace{1cm}} = 10$

15. $14 \div \underline{\hspace{1cm}} = 2$

16. $6 \div \underline{\hspace{1cm}} = 2$

17. $4 \div \underline{\hspace{1cm}} = 2$

18. $18 \div \underline{\hspace{1cm}} = 2$

19. $14 \div \underline{\hspace{1cm}} = 7$

20. $6 \div \underline{\hspace{1cm}} = 2$

21. $8 \div \underline{\hspace{1cm}} = 4$

22. $10 \div \underline{\hspace{1cm}} = 2$

23. $20 \div \underline{\hspace{1cm}} = 5$

24. $15 \div \underline{\hspace{1cm}} = 5$

25. $16 \div \underline{\hspace{1cm}} = 8$

26. $9 \div \underline{\hspace{1cm}} = 3$

27. $18 \div \underline{\hspace{1cm}} = 2$

28. $8 \div \underline{\hspace{1cm}} = 2$

29. $18 \div \underline{\hspace{1cm}} = 2$

30. $20 \div \underline{\hspace{1cm}} = 5$

31. $9 \div \underline{\hspace{1cm}} = 3$

32. $20 \div \underline{\hspace{1cm}} = 10$

33. $8 \div \underline{\hspace{1cm}} = 2$

34. $10 \div \underline{\hspace{1cm}} = 2$

35. $16 \div \underline{\hspace{1cm}} = 4$

36. $18 \div \underline{\hspace{1cm}} = 6$

37. $16 \div \underline{\hspace{1cm}} = 8$

38. $12 \div \underline{\hspace{1cm}} = 3$

39. $10 \div \underline{\hspace{1cm}} = 5$

40. $16 \div \underline{\hspace{1cm}} = 4$

41. $15 \div \underline{\hspace{1cm}} = 3$

42. $16 \div \underline{\hspace{1cm}} = 2$

43. $9 \div \underline{\hspace{1cm}} = 3$

44. $20 \div \underline{\hspace{1cm}} = 10$

45. $14 \div \underline{\hspace{1cm}} = 2$

46. $12 \div \underline{\hspace{1cm}} = 6$

47.	18 ÷ _____ = 2
48.	8 ÷ _____ = 2
49.	16 ÷ _____ = 2
50.	18 ÷ _____ = 3
1.	Ten ÷ Two = _____
2.	Six ÷ Three = _____
3.	Twelve ÷ Four = _____
4.	Ten ÷ Five = _____
5.	Six ÷ Two = _____
6.	Twenty ÷ Four = _____
7.	Eight ÷ Two = _____
8.	Twelve ÷ Four = _____
9.	Fourteen ÷ Seven = _____
10.	Twenty ÷ Five = _____
1.	Ten ÷ Two = Five
2.	Six ÷ Three = Two
3.	Twelve ÷ Four = Three
4.	Ten ÷ Five = Two

5.	Six ÷ Two = Three
6.	Twenty ÷ Four = Five
7.	Eight ÷ Two = Four
8.	Twelve ÷ Four = Three
9.	Fourteen ÷ Seven = Two
10.	Twenty ÷ Five = Four
1.	_____ ÷ Two = Five
2.	_____ ÷ Three = Two
3.	_____ ÷ Four = Three
4.	_____ ÷ Five = Two
5.	_____ ÷ Two = Three
6.	_____ ÷ Four = Five
7.	_____ ÷ Two = Four
8.	_____ ÷ Four = Three
9.	_____ ÷ Seven = Two
10.	_____ ÷ Five = Four
1.	Ten ÷ _____ = Five
2.	Six ÷ _____ = Two

3.	Twelve ÷ _____ = Three
4.	Ten ÷ _____ = Two
5.	Six ÷ _____ = Three
6.	Twenty ÷ _____ = Five
7.	Eight ÷ _____ = Four
8.	Twelve ÷ _____ = Three
9.	Fourteen ÷ _____ = Two
10.	Twenty ÷ _____ = Four

RENT-A-BIKES ON BROADWAY NEAR WORLD TRADE CENTER IN NYC

CHAPTER 2: K-2 MATH EXERCISES

CHAPTER 3: K-3 MATH EXERCISES

ADDITIONS

1.	65 + 62 = _____
2.	98 + 73 = _____
3.	53 + 15 = _____
4.	42 + 11 = _____
5.	26 + 14 = _____
6.	81 + 36 = _____
7.	90 + 9 = _____
8.	71 + 38 = _____
9.	74 + 24 = _____
10.	73 + 30 = _____
11.	59 + 40 = _____
12.	30 + 80 = _____
13.	36 + 20 = _____
14.	29 + 100 = _____

15.   61 + 34 = _____

16.   88 + 56 = _____

17.   50 + 3 = _____

18.   23 + 22 = _____

19.   69 + 88 = _____

20.   88 + 57 = _____

21.   83 + 87 = _____

22.   45 + 29 = _____

23.   4 + 84 = _____

24.   7 + 34 = _____

25.   63 + 25 = _____

26.   48 + 41 = _____

27.   70 + 17 = _____

28.   40 + 99 = _____

29.   29 + 29 = _____

30.   91 + 89 = _____

31.   17 + 34 = _____

32.   34 + 74 = _____

33.   76 + 22 = _____

34.   6 + 8 = _____

35.   14 + 54 = _____

36.   63 + 25 = _____

37.   13 + 73 = _____

38.   25 + 86 = _____

39.   39 + 25 = _____

40.   67 + 5 = _____

41.   21 + 69 = _____

42.   52 + 9 = _____

43.   17 + 23 = _____

44.   18 + 28 = _____

45.   14 + 42 = _____

46.   31 + 60 = _____

47.   52 + 89 = _____

48.   14 + 61 = _____

49.   21 + 33 = _____

50.   54 + 71 = _____

51.   89 + 11 = ____

52.   82 + 35 = ____

53.   80 + 26 = ____

54.   36 + 64 = ____

55.   59 + 39 = ____

56.   39 + 30 = ____

57.   64 + 82 = ____

58.   34 + 93 = ____

59.   18 + 51 = ____

60.   99 + 76 = ____

61.   43 + 47 = ____

62.   77 + 89 = ____

63.   71 + 29 = ____

64.   39 + 47 = ____

65.   26 + 32 = ____

66.   10 + 93 = ____

67.   33 + 17 = ____

68.   37 + 68 = ____

69.	97 + 38 = _____
70.	75 + 12 = _____
71.	60 + 89 = _____
72.	9 + 51 = _____
73.	85 + 23 = _____
74.	96 + 29 = _____
75.	74 + 43 = _____
76.	80 + 83 = _____
77.	4 + 85 = _____
78.	32 + 68 = _____
79.	44 + 53 = _____
80.	83 + 38 = _____
81.	50 + 96 = _____
82.	65 + 14 = _____
83.	64 + 63 = _____
84.	61 + 70 = _____
85.	16 + 10 = _____
86.	62 + 46 = _____

87.   59 + 66 = ____

88.   40 + 98 = ____

89.   9 + 16 = ____

90.   7 + 75 = ____

91.   81 + 99 = ____

92.   53 + 94 = ____

93.   25 + 60 = ____

94.   55 + 36 = ____

95.   59 + 85 = ____

96.   55 + 48 = ____

97.   22 + 49 = ____

98.   49 + 18 = ____

99.   34 + 50 = ____

100.   37 + 41 = ____

1.   65 + 62 = 127

2.   98 + 73 = 171

3.   53 + 15 = 68

4.   42 + 11 = 53

5.	26 + 14 = 40
6.	81 + 36 = 117
7.	90 + 9 = 99
8.	71 + 38 = 109
9.	74 + 24 = 98
10.	73 + 30 = 103
11.	59 + 40 = 99
12.	30 + 80 = 110
13.	36 + 20 = 56
14.	29 + 100 = 129
15.	61 + 34 = 95
16.	88 + 56 = 144
17.	50 + 3 = 53
18.	23 + 22 = 45
19.	69 + 88 = 157
20.	88 + 57 = 145
21.	83 + 87 = 170
22.	45 + 29 = 74

23.	4 + 84 = 88
24.	7 + 34 = 41
25.	63 + 25 = 88
26.	48 + 41 = 89
27.	70 + 17 = 87
28.	40 + 99 = 139
29.	29 + 29 = 58
30.	91 + 89 = 180
31.	17 + 34 = 51
32.	34 + 74 = 108
33.	76 + 22 = 98
34.	6 + 8 = 14
35.	14 + 54 = 68
36.	63 + 25 = 88
37.	13 + 73 = 86
38.	25 + 86 = 111
39.	39 + 25 = 64
40.	67 + 5 = 72

41.	21 + 69 = 90
42.	52 + 9 = 61
43.	17 + 23 = 40
44.	18 + 28 = 46
45.	14 + 42 = 56
46.	31 + 60 = 91
47.	52 + 89 = 141
48.	14 + 61 = 75
49.	21 + 33 = 54
50.	54 + 71 = 125
51.	89 + 11 = 100
52.	82 + 35 = 117
53.	80 + 26 = 106
54.	36 + 64 = 100
55.	59 + 39 = 98
56.	39 + 30 = 69
57.	64 + 82 = 146
58.	34 + 93 = 127

59.	18 + 51 = 69
60.	99 + 76 = 175
61.	43 + 47 = 90
62.	77 + 89 = 166
63.	71 + 29 = 100
64.	39 + 47 = 86
65.	26 + 32 = 58
66.	10 + 93 = 103
67.	33 + 17 = 50
68.	37 + 68 = 105
69.	97 + 38 = 135
70.	75 + 12 = 87
71.	60 + 89 = 149
72.	9 + 51 = 60
73.	85 + 23 = 108
74.	96 + 29 = 125
75.	74 + 43 = 117
76.	80 + 83 = 163

77.	$4 + 85 = 89$
78.	$32 + 68 = 100$
79.	$44 + 53 = 97$
80.	$83 + 38 = 121$
81.	$50 + 96 = 146$
82.	$65 + 14 = 79$
83.	$64 + 63 = 127$
84.	$61 + 70 = 131$
85.	$16 + 10 = 26$
86.	$62 + 46 = 108$
87.	$59 + 66 = 125$
88.	$40 + 98 = 138$
89.	$9 + 16 = 25$
90.	$7 + 75 = 82$
91.	$81 + 99 = 180$
92.	$53 + 94 = 147$
93.	$25 + 60 = 85$
94.	$55 + 36 = 91$

95.	59 + 85 = 144
96.	55 + 48 = 103
97.	22 + 49 = 71
98.	49 + 18 = 67
99.	34 + 50 = 84
100.	37 + 41 = 78
1.	____ + 62 = 127
2.	____ + 73 = 171
3.	____ + 15 = 68
4.	____ + 11 = 53
5.	____ + 14 = 40
6.	____ + 36 = 117
7.	____ + 9 = 99
8.	____ + 38 = 109
9.	____ + 24 = 98
10.	____ + 30 = 103
11.	____ + 40 = 99
12.	____ + 80 = 110

13.   _____ + 20 = 56

14.   _____ + 100 = 129

15.   _____ + 34 = 95

16.   _____ + 56 = 144

17.   _____ + 3 = 53

18.   _____ + 22 = 45

19.   _____ + 88 = 157

20.   _____ + 57 = 145

21.   _____ + 87 = 170

22.   _____ + 29 = 74

23.   _____ + 84 = 88

24.   _____ + 34 = 41

25.   _____ + 25 = 88

26.   _____ + 41 = 89

27.   _____ + 17 = 87

28.   _____ + 99 = 139

29.   _____ + 29 = 58

30.   _____ + 89 = 180

CHAPTER 3: K-3 MATH EXERCISES

31.   _____ + 34 = 51

32.   _____ + 74 = 108

33.   _____ + 22 = 98

34.   _____ + 8 = 14

35.   _____ + 54 = 68

36.   _____ + 25 = 88

37.   _____ + 73 = 86

38.   _____ + 86 = 111

39.   _____ + 25 = 64

40.   _____ + 5 = 72

41.   _____ + 69 = 90

42.   _____ + 9 = 61

43.   _____ + 23 = 40

44.   _____ + 28 = 46

45.   _____ + 42 = 56

46.   _____ + 60 = 91

47.   _____ + 89 = 141

48.   _____ + 61 = 75

49. _____ + 33 = 54

50. _____ + 71 = 125

51. _____ + 11 = 100

52. _____ + 35 = 117

53. _____ + 26 = 106

54. _____ + 64 = 100

55. _____ + 39 = 98

56. _____ + 30 = 69

57. _____ + 82 = 146

58. _____ + 93 = 127

59. _____ + 51 = 69

60. _____ + 76 = 175

61. _____ + 47 = 90

62. _____ + 89 = 166

63. _____ + 29 = 100

64. _____ + 47 = 86

65. _____ + 32 = 58

66. _____ + 93 = 103

67.   _____ + 17 = 50

68.   _____ + 68 = 105

69.   _____ + 38 = 135

70.   _____ + 12 = 87

71.   _____ + 89 = 149

72.   _____ + 51 = 60

73.   _____ + 23 = 108

74.   _____ + 29 = 125

75.   _____ + 43 = 117

76.   _____ + 83 = 163

77.   _____ + 85 = 89

78.   _____ + 68 = 100

79.   _____ + 53 = 97

80.   _____ + 38 = 121

81.   _____ + 96 = 146

82.   _____ + 14 = 79

83.   _____ + 63 = 127

84.   _____ + 70 = 131

85.   _____ + 10 = 26

86.   _____ + 46 = 108

87.   _____ + 66 = 125

88.   _____ + 98 = 138

89.   _____ + 16 = 25

90.   _____ + 75 = 82

91.   _____ + 99 = 180

92.   _____ + 94 = 147

93.   _____ + 60 = 85

94.   _____ + 36 = 91

95.   _____ + 85 = 144

96.   _____ + 48 = 103

97.   _____ + 49 = 71

98.   _____ + 18 = 67

99.   _____ + 50 = 84

100.   _____ + 41 = 78

1.   65 + _____ = 127

2.   98 + _____ = 171

3.   53 + ____ = 68

4.   42 + ____ = 53

5.   26 + ____ = 40

6.   81 + ____ = 117

7.   90 + ____ = 99

8.   71 + ____ = 109

9.   74 + ____ = 98

10.   73 + ____ = 103

11.   59 + ____ = 99

12.   30 + ____ = 110

13.   36 + ____ = 56

14.   29 + ____ = 129

15.   61 + ____ = 95

16.   88 + ____ = 144

17.   50 + ____ = 53

18.   23 + ____ = 45

19.   69 + ____ = 157

20.   88 + ____ = 145

21.   83 + _____ = 170

22.   45 + _____ = 74

23.   4 + _____ = 88

24.   7 + _____ = 41

25.   63 + _____ = 88

26.   48 + _____ = 89

27.   70 + _____ = 87

28.   40 + _____ = 139

29.   29 + _____ = 58

30.   91 + _____ = 180

31.   17 + _____ = 51

32.   34 + _____ = 108

33.   76 + _____ = 98

34.   6 + _____ = 14

35.   14 + _____ = 68

36.   63 + _____ = 88

37.   13 + _____ = 86

38.   25 + _____ = 111

39.　　　　$39 + \underline{\quad} = 64$

40.　　　　$67 + \underline{\quad} = 72$

41.　　　　$21 + \underline{\quad} = 90$

42.　　　　$52 + \underline{\quad} = 61$

43.　　　　$17 + \underline{\quad} = 40$

44.　　　　$18 + \underline{\quad} = 46$

45.　　　　$14 + \underline{\quad} = 56$

46.　　　　$31 + \underline{\quad} = 91$

47.　　　　$52 + \underline{\quad} = 141$

48.　　　　$14 + \underline{\quad} = 75$

49.　　　　$21 + \underline{\quad} = 54$

50.　　　　$54 + \underline{\quad} = 125$

51.　　　　$89 + \underline{\quad} = 100$

52.　　　　$82 + \underline{\quad} = 117$

53.　　　　$80 + \underline{\quad} = 106$

54.　　　　$36 + \underline{\quad} = 100$

55.　　　　$59 + \underline{\quad} = 98$

56.　　　　$39 + \underline{\quad} = 69$

57.	64 + ____ = 146
58.	34 + ____ = 127
59.	18 + ____ = 69
60.	99 + ____ = 175
61.	43 + ____ = 90
62.	77 + ____ = 166
63.	71 + ____ = 100
64.	39 + ____ = 86
65.	26 + ____ = 58
66.	10 + ____ = 103
67.	33 + ____ = 50
68.	37 + ____ = 105
69.	97 + ____ = 135
70.	75 + ____ = 87
71.	60 + ____ = 149
72.	9 + ____ = 60
73.	85 + ____ = 108
74.	96 + ____ = 125

75. $74 + \underline{} = 117$

76. $80 + \underline{} = 163$

77. $4 + \underline{} = 89$

78. $32 + \underline{} = 100$

79. $44 + \underline{} = 97$

80. $83 + \underline{} = 121$

81. $50 + \underline{} = 146$

82. $65 + \underline{} = 79$

83. $64 + \underline{} = 127$

84. $61 + \underline{} = 131$

85. $16 + \underline{} = 26$

86. $62 + \underline{} = 108$

87. $59 + \underline{} = 125$

88. $40 + \underline{} = 138$

89. $9 + \underline{} = 25$

90. $7 + \underline{} = 82$

91. $81 + \underline{} = 180$

92. $53 + \underline{} = 147$

93.	25 + ____ = 85
94.	55 + ____ = 91
95.	59 + ____ = 144
96.	55 + ____ = 103
97.	22 + ____ = 71
98.	49 + ____ = 67
99.	34 + ____ = 84
100.	37 + ____ = 78
1.	Sixty-Five + Sixty-Two = ____
2.	Ninety-Eight + Seventy-Three = ____
3.	Fifty-Three + Fifteen = ____
4.	Forty-Two + Eleven = ____
5.	Twenty-Six + Fourteen = ____
6.	Eighty-One + Thirty-Six = ____
7.	Ninety + Nine = ____
8.	Seventy-One + Thirty-Eight = ____
9.	Seventy-Four + Twenty-Four = ____
10.	Seventy-Three + Thirty = ____

11.	Fifty-Nine + Forty = ____
12.	Thirty + Eighty = ____
13.	Thirty-Six + Twenty = ____
14.	Twenty-Nine + One Hundred = ____
15.	Sixty-One + Thirty-Four = ____
16.	Eighty-Eight + Fifty-Six = ____
17.	Fifty + Three = ____
18.	Twenty-Three + Twenty-Two = ____
19.	Sixty-Nine + Eighty-Eight = ____
20.	Eighty-Eight + Fifty-Seven = ____
1.	Sixty-Five + Sixty-Two = One Hundred Twenty-Seven
2.	Ninety-Eight + Seventy-Three = One Hundred Seventy-One
3.	Fifty-Three + Fifteen = Sixty-Eight
4.	Forty-Two + Eleven = Fifty-Three
5.	Twenty-Six + Fourteen = Forty
6.	Eighty-One + Thirty-Six = One Hundred Seventeen

7.	Ninety + Nine = Ninety-Nine
8.	Seventy-One + Thirty-Eight = One Hundred Nine
9.	Seventy-Four + Twenty-Four = Ninety-Eight
10.	Seventy-Three + Thirty = One Hundred Three
11.	Fifty-Nine + Forty = Ninety-Nine
12.	Thirty + Eighty = One Hundred Ten
13.	Thirty-Six + Twenty = Fifty-Six
14.	Twenty-Nine + One Hundred = One Hundred Twenty-Nine
15.	Sixty-One + Thirty-Four = Ninety-Five
16.	Eighty-Eight + Fifty-Six = One Hundred Forty-Four
17.	Fifty + Three = Fifty-Three
18.	Twenty-Three + Twenty-Two = Forty-Five
19.	Sixty-Nine + Eighty-Eight = One Hundred Fifty-Seven
20.	Eighty-Eight + Fifty-Seven = One Hundred Forty-Five
1.	_____ + Sixty-Two = One Hundred Twenty-Seven
2.	_____ + Seventy-Three = One Hundred Seventy-

	One
3.	_____ + Fifteen = Sixty-Eight
4.	_____ + Eleven = Fifty-Three
5.	_____ + Fourteen = Forty
6.	_____ + Thirty-Six = One Hundred Seventeen
7.	_____ + Nine = Ninety-Nine
8.	_____ + Thirty-Eight = One Hundred Nine
9.	_____ + Twenty-Four = Ninety-Eight
10.	_____ + Thirty = One Hundred Three
11.	_____ + Forty = Ninety-Nine
12.	_____ + Eighty = One Hundred Ten
13.	_____ + Twenty = Fifty-Six
14.	_____ + One Hundred = One Hundred Twenty-Nine
15.	_____ + Thirty-Four = Ninety-Five
16.	_____ + Fifty-Six = One Hundred Forty-Four
17.	_____ + Three = Fifty-Three
18.	_____ + Twenty-Two = Forty-Five
19.	_____ + Eighty-Eight = One Hundred Fifty-Seven

20.	____ + Fifty-Seven = One Hundred Forty-Five
1.	Sixty-Five + ____ = One Hundred Twenty-Seven
2.	Ninety-Eight + ____ = One Hundred Seventy-One
3.	Fifty-Three + ____ = Sixty-Eight
4.	Forty-Two + ____ = Fifty-Three
5.	Twenty-Six + ____ = Forty
6.	Eighty-One + ____ = One Hundred Seventeen
7.	Ninety + ____ = Ninety-Nine
8.	Seventy-One + ____ = One Hundred Nine
9.	Seventy-Four + ____ = Ninety-Eight
10.	Seventy-Three + ____ = One Hundred Three
11.	Fifty-Nine + ____ = Ninety-Nine
12.	Thirty + ____ = One Hundred Ten
13.	Thirty-Six + ____ = Fifty-Six
14.	Twenty-Nine + ____ = One Hundred Twenty-Nine
15.	Sixty-One + ____ = Ninety-Five
16.	Eighty-Eight + ____ = One Hundred Forty-Four

CHRISTOPHER COLUMBUS'S STATUES IN MADRID AND NYC

CHAPTER 3: K-3 MATH EXERCISES

COOKING ON COLUMBUS'S SHIP PINTA IN 1492

CHAPTER 3: K-3 MATH EXERCISES

SUBTRACTIONS

1.   85 - 70 = _____

2.   86 - 75 = _____

3.   77 - 66 = _____

4.   75 - 47 = _____

5.   100 - 9 = _____

6.   70 - 10 = _____

7.   55 - 54 = _____

8.   61 - 24 = _____

9.   88 - 19 = _____

10.   81 - 67 = _____

11.   60 - 14 = _____

12.   25 - 16 = _____

13.   57 - 26 = _____

14.   77 - 7 = _____

15.   53 - 44 = _____

16.   90 - 42 = _____

17.   72 - 35 = ____

18.   77 - 25 = ____

19.   42 - 8 = ____

20.   85 - 9 = ____

21.   23 - 16 = ____

22.   58 - 18 = ____

23.   95 - 67 = ____

24.   81 - 13 = ____

25.   89 - 70 = ____

26.   92 - 87 = ____

27.   66 - 19 = ____

28.   68 - 14 = ____

29.   37 - 22 = ____

30.   100 - 94 = ____

31.   26 - 7 = ____

32.   14 - 3 = ____

33.   87 - 74 = ____

34.   51 - 21 = ____

35.   24 - 20 = _____

36.   71 - 37 = _____

37.   47 - 4 = _____

38.   38 - 24 = _____

39.   80 - 51 = _____

40.   96 - 90 = _____

41.   73 - 48 = _____

42.   86 - 42 = _____

43.   43 - 8 = _____

44.   73 - 14 = _____

45.   49 - 36 = _____

46.   75 - 24 = _____

47.   91 - 62 = _____

48.   59 - 10 = _____

49.   44 - 11 = _____

50.   82 - 40 = _____

51.   56 - 36 = _____

52.   51 - 39 = _____

CHAPTER 3: K-3 MATH EXERCISES

53.   93 - 73 = ____

54.   94 - 37 = ____

55.   77 - 12 = ____

56.   78 - 5 = ____

57.   22 - 13 = ____

58.   39 - 25 = ____

59.   78 - 32 = ____

60.   100 - 3 = ____

61.   83 - 42 = ____

62.   99 - 80 = ____

63.   62 - 5 = ____

64.   73 - 51 = ____

65.   84 - 49 = ____

66.   69 - 32 = ____

67.   89 - 12 = ____

68.   95 - 41 = ____

69.   25 - 8 = ____

70.   63 - 34 = ____

71.	56 - 14 = _____
72.	93 - 12 = _____
73.	61 - 39 = _____
74.	80 - 3 = _____
75.	60 - 24 = _____
76.	71 - 18 = _____
77.	97 - 82 = _____
78.	93 - 71 = _____
79.	100 - 60 = _____
80.	37 - 12 = _____
81.	90 - 47 = _____
82.	85 - 55 = _____
83.	66 - 37 = _____
84.	62 - 18 = _____
85.	94 - 13 = _____
86.	100 - 33 = _____
87.	83 - 73 = _____
88.	96 - 24 = _____

89.	55 - 11 = ____
90.	82 - 73 = ____
91.	44 - 4 = ____
92.	83 - 18 = ____
93.	86 - 59 = ____
94.	75 - 65 = ____
95.	78 - 38 = ____
96.	93 - 41 = ____
97.	86 - 60 = ____
98.	25 - 4 = ____
99.	75 - 45 = ____
100.	77 - 10 = ____
1.	85 - 70 = 15
2.	86 - 75 = 11
3.	77 - 66 = 11
4.	75 - 47 = 28
5.	100 - 9 = 91
6.	70 - 10 = 60

CHAPTER 3: K-3 MATH EXERCISES

7.	55 - 54 = 1
8.	61 - 24 = 37
9.	88 - 19 = 69
10.	81 - 67 = 14
11.	60 - 14 = 46
12.	25 - 16 = 9
13.	57 - 26 = 31
14.	77 - 7 = 70
15.	53 - 44 = 9
16.	90 - 42 = 48
17.	72 - 35 = 37
18.	77 - 25 = 52
19.	42 - 8 = 34
20.	85 - 9 = 76
21.	23 - 16 = 7
22.	58 - 18 = 40
23.	95 - 67 = 28
24.	81 - 13 = 68

25.	89 - 70 = 19
26.	92 - 87 = 5
27.	66 - 19 = 47
28.	68 - 14 = 54
29.	37 - 22 = 15
30.	100 - 94 = 6
31.	26 - 7 = 19
32.	14 - 3 = 11
33.	87 - 74 = 13
34.	51 - 21 = 30
35.	24 - 20 = 4
36.	71 - 37 = 34
37.	47 - 4 = 43
38.	38 - 24 = 14
39.	80 - 51 = 29
40.	96 - 90 = 6
41.	73 - 48 = 25
42.	86 - 42 = 44

43.	43 - 8 = 35
44.	73 - 14 = 59
45.	49 - 36 = 13
46.	75 - 24 = 51
47.	91 - 62 = 29
48.	59 - 10 = 49
49.	44 - 11 = 33
50.	82 - 40 = 42
51.	56 - 36 = 20
52.	51 - 39 = 12
53.	93 - 73 = 20
54.	94 - 37 = 57
55.	77 - 12 = 65
56.	78 - 5 = 73
57.	22 - 13 = 9
58.	39 - 25 = 14
59.	78 - 32 = 46
60.	100 - 3 = 97

61.	83 - 42 = 41
62.	99 - 80 = 19
63.	62 - 5 = 57
64.	73 - 51 = 22
65.	84 - 49 = 35
66.	69 - 32 = 37
67.	89 - 12 = 77
68.	95 - 41 = 54
69.	25 - 8 = 17
70.	63 - 34 = 29
71.	56 - 14 = 42
72.	93 - 12 = 81
73.	61 - 39 = 22
74.	80 - 3 = 77
75.	60 - 24 = 36
76.	71 - 18 = 53
77.	97 - 82 = 15
78.	93 - 71 = 22

79.	100 - 60 = 40
80.	37 - 12 = 25
81.	90 - 47 = 43
82.	85 - 55 = 30
83.	66 - 37 = 29
84.	62 - 18 = 44
85.	94 - 13 = 81
86.	100 - 33 = 67
87.	83 - 73 = 10
88.	96 - 24 = 72
89.	55 - 11 = 44
90.	82 - 73 = 9
91.	44 - 4 = 40
92.	83 - 18 = 65
93.	86 - 59 = 27
94.	75 - 65 = 10
95.	78 - 38 = 40
96.	93 - 41 = 52

97.	86 - 60 = 26
98.	25 - 4 = 21
99.	75 - 45 = 30
100.	77 - 10 = 67
1.	____ - 70 = 15
2.	____ - 75 = 11
3.	____ - 66 = 11
4.	____ - 47 = 28
5.	____ - 9 = 91
6.	____ - 10 = 60
7.	____ - 54 = 1
8.	____ - 24 = 37
9.	____ - 19 = 69
10.	____ - 67 = 14
11.	____ - 14 = 46
12.	____ - 16 = 9
13.	____ - 26 = 31
14.	____ - 7 = 70

15. ____ - 44 = 9

16. ____ - 42 = 48

17. ____ - 35 = 37

18. ____ - 25 = 52

19. ____ - 8 = 34

20. ____ - 9 = 76

21. ____ - 16 = 7

22. ____ - 18 = 40

23. ____ - 67 = 28

24. ____ - 13 = 68

25. ____ - 70 = 19

26. ____ - 87 = 5

27. ____ - 19 = 47

28. ____ - 14 = 54

29. ____ - 22 = 15

30. ____ - 94 = 6

31. ____ - 7 = 19

32. ____ - 3 = 11

33.   ____ - 74 = 13

34.   ____ - 21 = 30

35.   ____ - 20 = 4

36.   ____ - 37 = 34

37.   ____ - 4 = 43

38.   ____ - 24 = 14

39.   ____ - 51 = 29

40.   ____ - 90 = 6

41.   ____ - 48 = 25

42.   ____ - 42 = 44

43.   ____ - 8 = 35

44.   ____ - 14 = 59

45.   ____ - 36 = 13

46.   ____ - 24 = 51

47.   ____ - 62 = 29

48.   ____ - 10 = 49

49.   ____ - 11 = 33

50.   ____ - 40 = 42

51. _____ - 36 = 20

52. _____ - 39 = 12

53. _____ - 73 = 20

54. _____ - 37 = 57

55. _____ - 12 = 65

56. _____ - 5 = 73

57. _____ - 13 = 9

58. _____ - 25 = 14

59. _____ - 32 = 46

60. _____ - 3 = 97

61. _____ - 42 = 41

62. _____ - 80 = 19

63. _____ - 5 = 57

64. _____ - 51 = 22

65. _____ - 49 = 35

66. _____ - 32 = 37

67. _____ - 12 = 77

68. _____ - 41 = 54

69.	_____ - 8 = 17
70.	_____ - 34 = 29
71.	_____ - 14 = 42
72.	_____ - 12 = 81
73.	_____ - 39 = 22
74.	_____ - 3 = 77
75.	_____ - 24 = 36
76.	_____ - 18 = 53
77.	_____ - 82 = 15
78.	_____ - 71 = 22
79.	_____ - 60 = 40
80.	_____ - 12 = 25
81.	_____ - 47 = 43
82.	_____ - 55 = 30
83.	_____ - 37 = 29
84.	_____ - 18 = 44
85.	_____ - 13 = 81
86.	_____ - 33 = 67

87.   ____ - 73 = 10

88.   ____ - 24 = 72

89.   ____ - 11 = 44

90.   ____ - 73 = 9

91.   ____ - 4 = 40

92.   ____ - 18 = 65

93.   ____ - 59 = 27

94.   ____ - 65 = 10

95.   ____ - 38 = 40

96.   ____ - 41 = 52

97.   ____ - 60 = 26

98.   ____ - 4 = 21

99.   ____ - 45 = 30

100.   ____ - 10 = 67

1.   85 - ____ = 15

2.   86 - ____ = 11

3.   77 - ____ = 11

4.   75 - ____ = 28

5.   100 - _____ = 91

6.   70 - _____ = 60

7.   55 - _____ = 1

8.   61 - _____ = 37

9.   88 - _____ = 69

10. 81 - _____ = 14

11. 60 - _____ = 46

12. 25 - _____ = 9

13. 57 - _____ = 31

14. 77 - _____ = 70

15. 53 - _____ = 9

16. 90 - _____ = 48

17. 72 - _____ = 37

18. 77 - _____ = 52

19. 42 - _____ = 34

20. 85 - _____ = 76

21. 23 - _____ = 7

22. 58 - _____ = 40

23.   95 - ____ = 28

24.   81 - ____ = 68

25.   89 - ____ = 19

26.   92 - ____ = 5

27.   66 - ____ = 47

28.   68 - ____ = 54

29.   37 - ____ = 15

30.   100 - ____ = 6

31.   26 - ____ = 19

32.   14 - ____ = 11

33.   87 - ____ = 13

34.   51 - ____ = 30

35.   24 - ____ = 4

36.   71 - ____ = 34

37.   47 - ____ = 43

38.   38 - ____ = 14

39.   80 - ____ = 29

40.   96 - ____ = 6

41.   73 - _____ = 25

42.   86 - _____ = 44

43.   43 - _____ = 35

44.   73 - _____ = 59

45.   49 - _____ = 13

46.   75 - _____ = 51

47.   91 - _____ = 29

48.   59 - _____ = 49

49.   44 - _____ = 33

50.   82 - _____ = 42

51.   56 - _____ = 20

52.   51 - _____ = 12

53.   93 - _____ = 20

54.   94 - _____ = 57

55.   77 - _____ = 65

56.   78 - _____ = 73

57.   22 - _____ = 9

58.   39 - _____ = 14

59.   78 - _____ = 46

60.   100 - _____ = 97

61.   83 - _____ = 41

62.   99 - _____ = 19

63.   62 - _____ = 57

64.   73 - _____ = 22

65.   84 - _____ = 35

66.   69 - _____ = 37

67.   89 - _____ = 77

68.   95 - _____ = 54

69.   25 - _____ = 17

70.   63 - _____ = 29

71.   56 - _____ = 42

72.   93 - _____ = 81

73.   61 - _____ = 22

74.   80 - _____ = 77

75.   60 - _____ = 36

76.   71 - _____ = 53

77.   97 - ____ = 15

78.   93 - ____ = 22

79.   100 - ____ = 40

80.   37 - ____ = 25

81.   90 - ____ = 43

82.   85 - ____ = 30

83.   66 - ____ = 29

84.   62 - ____ = 44

85.   94 - ____ = 81

86.   100 - ____ = 67

87.   83 - ____ = 10

88.   96 - ____ = 72

89.   55 - ____ = 44

90.   82 - ____ = 9

91.   44 - ____ = 40

92.   83 - ____ = 65

93.   86 - ____ = 27

94.   75 - ____ = 10

95.	78 - ____ = 40
96.	93 - ____ = 52
97.	86 - ____ = 26
98.	25 - ____ = 21
99.	75 - ____ = 30
100.	77 - ____ = 67
1.	Eighty-Five - Seventy = ____
2.	Eighty-Six - Seventy-Five = ____
3.	Seventy-Seven - Sixty-Six = ____
4.	Seventy-Five - Forty-Seven = ____
5.	One Hundred - Nine = ____
6.	Seventy - Ten = ____
7.	Fifty-Five - Fifty-Four = ____
8.	Sixty-One - Twenty-Four = ____
9.	Eighty-Eight - Nineteen = ____
10.	Eighty-One - Sixty-Seven = ____
11.	Sixty - Fourteen = ____
12.	Twenty-Five - Sixteen = ____

13.	Fifty-Seven - Twenty-Six = _____
14.	Seventy-Seven - Seven = _____
15.	Fifty-Three - Forty-Four = _____
16.	Ninety - Forty-Two = _____
17.	Seventy-Two - Thirty-Five = _____
18.	Seventy-Seven - Twenty-Five = _____
19.	Forty-Two - Eight = _____
20.	Eighty-Five - Nine = _____
1.	Eighty-Five - Seventy = Fifteen
2.	Eighty-Six - Seventy-Five = Eleven
3.	Seventy-Seven - Sixty-Six = Eleven
4.	Seventy-Five - Forty-Seven = Twenty-Eight
5.	One Hundred - Nine = Ninety-One
6.	Seventy - Ten = Sixty
7.	Fifty-Five - Fifty-Four = One
8.	Sixty-One - Twenty-Four = Thirty-Seven
9.	Eighty-Eight - Nineteen = Sixty-Nine

10.	Eighty-One - Sixty-Seven = Fourteen
11.	Sixty - Fourteen = Forty-Six
12.	Twenty-Five - Sixteen = Nine
13.	Fifty-Seven - Twenty-Six = Thirty-One
14.	Seventy-Seven - Seven = Seventy
15.	Fifty-Three - Forty-Four = Nine
16.	Ninety - Forty-Two = Forty-Eight
17.	Seventy-Two - Thirty-Five = Thirty-Seven
18.	Seventy-Seven - Twenty-Five = Fifty-Two
19.	Forty-Two - Eight = Thirty-Four
20.	Eighty-Five - Nine = Seventy-Six
1.	_____ - Seventy = Fifteen
2.	_____ - Seventy-Five = Eleven
3.	_____ - Sixty-Six = Eleven
4.	_____ - Forty-Seven = Twenty-Eight
5.	_____ - Nine = Ninety-One
6.	_____ - Ten = Sixty

7.	_____ - Fifty-Four = One
8.	_____ - Twenty-Four = Thirty-Seven
9.	_____ - Nineteen = Sixty-Nine
10.	_____ - Sixty-Seven = Fourteen
11.	_____ - Fourteen = Forty-Six
12.	_____ - Sixteen = Nine
13.	_____ - Twenty-Six = Thirty-One
14.	_____ - Seven = Seventy
15.	_____ - Forty-Four = Nine
16.	_____ - Forty-Two = Forty-Eight
17.	_____ - Thirty-Five = Thirty-Seven
18.	_____ - Twenty-Five = Fifty-Two
19.	_____ - Eight = Thirty-Four
20.	_____ - Nine = Seventy-Six
1.	Eighty-Five - _____ = Fifteen
2.	Eighty-Six - _____ = Eleven
3.	Seventy-Seven - _____ = Eleven
4.	Seventy-Five - _____ = Twenty-Eight

5.	One Hundred - _____ = Ninety-One
6.	Seventy - _____ = Sixty
7.	Fifty-Five - _____ = One
8.	Sixty-One - _____ = Thirty-Seven
9.	Eighty-Eight - _____ = Sixty-Nine
10.	Eighty-One - _____ = Fourteen
11.	Sixty - _____ = Forty-Six
12.	Twenty-Five - _____ = Nine
13.	Fifty-Seven - _____ = Thirty-One
14.	Seventy-Seven - _____ = Seventy
15.	Fifty-Three - _____ = Nine
16.	Ninety - _____ = Forty-Eight
17.	Seventy-Two - _____ = Thirty-Seven
18.	Seventy-Seven - _____ = Fifty-Two
19.	Forty-Two - _____ = Thirty-Four
20.	Eighty-Five - _____ = Seventy-Six

3D SCULPTURE ON PARK AVENUE IN NEW YORK CITY

CHAPTER 3: K-3 MATH EXERCISES

MULTIPLICATIONS

1.   14 x 14 = _____

2.   12 x 6 = _____

3.   10 x 7 = _____

4.   15 x 15 = _____

5.   11 x 9 = _____

6.   4 x 4 = _____

7.   7 x 10 = _____

8.   13 x 6 = _____

9.   11 x 7 = _____

10.   14 x 9 = _____

11.   15 x 3 = _____

12.   3 x 14 = _____

13.   9 x 15 = _____

14.   13 x 5 = _____

15.   6 x 3 = _____

16.   4 x 15 = _____

17.   10 x 9 = _____

18.   3 x 10 = _____

19.   12 x 6 = _____

20.   6 x 10 = _____

21.   3 x 15 = _____

22.   8 x 11 = _____

23.   12 x 12 = _____

24.   9 x 11 = _____

25.   10 x 5 = _____

26.   6 x 10 = _____

27.   11 x 15 = _____

28.   9 x 8 = _____

29.   4 x 12 = _____

30.   7 x 13 = _____

31.   9 x 11 = _____

32.   15 x 8 = _____

33.   **5 x 13 = _____**

34.   **10 x 11 = _____**

35.   **12 x 6 = _____**

36.   **11 x 8 = _____**

37.   **9 x 12 = _____**

38.   **4 x 13 = _____**

39.   **9 x 10 = _____**

40.   **7 x 7 = _____**

41.   **11 x 8 = _____**

42.   **8 x 3 = _____**

43.   **3 x 6 = _____**

44.   **7 x 10 = _____**

45.   **12 x 3 = _____**

46.   **3 x 8 = _____**

47.   **4 x 5 = _____**

48.   **15 x 7 = _____**

49.   **12 x 11 = _____**

50.   13 x 6 = _____

1.   14 x 14 = 196

2.   12 x 6 = 72

3.   10 x 7 = 70

4.   15 x 15 = 225

5.   11 x 9 = 99

6.   4 x 4 = 16

7.   7 x 10 = 70

8.   13 x 6 = 78

9.   11 x 7 = 77

10.   14 x 9 = 126

11.   15 x 3 = 45

12.   3 x 14 = 42

13.   9 x 15 = 135

14.   13 x 5 = 65

15.   6 x 3 = 18

16.   4 x 15 = 60

17.	10 x 9 = 90
18.	3 x 10 = 30
19.	12 x 6 = 72
20.	6 x 10 = 60
21.	3 x 15 = 45
22.	8 x 11 = 88
23.	12 x 12 = 144
24.	9 x 11 = 99
25.	10 x 5 = 50
26.	6 x 10 = 60
27.	11 x 15 = 165
28.	9 x 8 = 72
29.	4 x 12 = 48
30.	7 x 13 = 91
31.	9 x 11 = 99
32.	15 x 8 = 120
33.	5 x 13 = 65

34.	10 x 11 = 110
35.	12 x 6 = 72
36.	11 x 8 = 88
37.	9 x 12 = 108
38.	4 x 13 = 52
39.	9 x 10 = 90
40.	7 x 7 = 49
41.	11 x 8 = 88
42.	8 x 3 = 24
43.	3 x 6 = 18
44.	7 x 10 = 70
45.	12 x 3 = 36
46.	3 x 8 = 24
47.	4 x 5 = 20
48.	15 x 7 = 105
49.	12 x 11 = 132
50.	13 x 6 = 78

1.   _____ x 14 = 196

2.   _____ x 6 = 72

3.   _____ x 7 = 70

4.   _____ x 15 = 225

5.   _____ x 9 = 99

6.   _____ x 4 = 16

7.   _____ x 10 = 70

8.   _____ x 6 = 78

9.   _____ x 7 = 77

10.   _____ x 9 = 126

11.   _____ x 3 = 45

12.   _____ x 14 = 42

13.   _____ x 15 = 135

14.   _____ x 5 = 65

15.   _____ x 3 = 18

16.   _____ x 15 = 60

17.   _____ x 9 = 90

18. ____ x 10 = 30

19. ____ x 6 = 72

20. ____ x 10 = 60

21. ____ x 15 = 45

22. ____ x 11 = 88

23. ____ x 12 = 144

24. ____ x 11 = 99

25. ____ x 5 = 50

26. ____ x 10 = 60

27. ____ x 15 = 165

28. ____ x 8 = 72

29. ____ x 12 = 48

30. ____ x 13 = 91

31. ____ x 11 = 99

32. ____ x 8 = 120

33. ____ x 13 = 65

34. ____ x 11 = 110

35. _____ x 6 = 72

36. _____ x 8 = 88

37. _____ x 12 = 108

38. _____ x 13 = 52

39. _____ x 10 = 90

40. _____ x 7 = 49

41. _____ x 8 = 88

42. _____ x 3 = 24

43. _____ x 6 = 18

44. _____ x 10 = 70

45. _____ x 3 = 36

46. _____ x 8 = 24

47. _____ x 5 = 20

48. _____ x 7 = 105

49. _____ x 11 = 132

50. _____ x 6 = 78

1. 14 x _____ = 196

2.   12 x _____ = 72

3.   10 x _____ = 70

4.   15 x _____ = 225

5.   11 x _____ = 99

6.   4 x _____ = 16

7.   7 x _____ = 70

8.   13 x _____ = 78

9.   11 x _____ = 77

10.   14 x _____ = 126

11.   15 x _____ = 45

12.   3 x _____ = 42

13.   9 x _____ = 135

14.   13 x _____ = 65

15.   6 x _____ = 18

16.   4 x _____ = 60

17.   10 x _____ = 90

18.   3 x _____ = 30

19.   12 x _____ = 72

20.   6 x _____ = 60

21.   3 x _____ = 45

22.   8 x _____ = 88

23.   12 x _____ = 144

24.   9 x _____ = 99

25.   10 x _____ = 50

26.   6 x _____ = 60

27.   11 x _____ = 165

28.   9 x _____ = 72

29.   4 x _____ = 48

30.   7 x _____ = 91

31.   9 x _____ = 99

32.   15 x _____ = 120

33.   5 x _____ = 65

34.   10 x _____ = 110

35.   12 x _____ = 72

36. 11 x _____ = 88

37. 9 x _____ = 108

38. 4 x _____ = 52

39. 9 x _____ = 90

40. 7 x _____ = 49

41. 11 x _____ = 88

42. 8 x _____ = 24

43. 3 x _____ = 18

44. 7 x _____ = 70

45. 12 x _____ = 36

46. 3 x _____ = 24

47. 4 x _____ = 20

48. 15 x _____ = 105

49. 12 x _____ = 132

50. 13 x _____ = 78

1. Fourteen x Fourteen = _____

2. Twelve x Six = _____

3.   Ten x Seven = _____

4.   Fifteen x Fifteen = _____

5.   Eleven x Nine = _____

6.   Four x Four = _____

7.   Seven x Ten = _____

8.   Thirteen x Six = _____

9.   Eleven x Seven = _____

10.   Fourteen x Nine = _____

11.   Fifteen x Three = _____

12.   Three x Fourteen = _____

13.   Nine x Fifteen = _____

14.   Thirteen x Five = _____

15.   Six x Three = _____

16.   Four x Fifteen = _____

17.   Ten x Nine = _____

18.   Three x Ten = _____

19.   Twelve x Six = _____

20.	Six x Ten = _____
1.	Fourteen x Fourteen = One Hundred Ninety-Six
2.	Twelve x Six = Seventy-Two
3.	Ten x Seven = Seventy
4.	Fifteen x Fifteen = Two Hundred Twenty-Five
5.	Eleven x Nine = Ninety-Nine
6.	Four x Four = Sixteen
7.	Seven x Ten = Seventy
8.	Thirteen x Six = Seventy-Eight
9.	Eleven x Seven = Seventy-Seven
10.	Fourteen x Nine = One Hundred Twenty-Six
11.	Fifteen x Three = Forty-Five
12.	Three x Fourteen = Forty-Two
13.	Nine x Fifteen = One Hundred Thirty-Five
14.	Thirteen x Five = Sixty-Five
15.	Six x Three = Eighteen
16.	Four x Fifteen = Sixty

17.	Ten x Nine = Ninety
18.	Three x Ten = Thirty
19.	Twelve x Six = Seventy-Two
20.	Six x Ten = Sixty
1.	_____ x Fourteen = One Hundred Ninety-Six
2.	_____ x Six = Seventy-Two
3.	_____ x Seven = Seventy
4.	_____ x Fifteen = Two Hundred Twenty-Five
5.	_____ x Nine = Ninety-Nine
6.	_____ x Four = Sixteen
7.	_____ x Ten = Seventy
8.	_____ x Six = Seventy-Eight
9.	_____ x Seven = Seventy-Seven
10.	_____ x Nine = One Hundred Twenty-Six
11.	_____ x Three = Forty-Five
12.	_____ x Fourteen = Forty-Two
13.	_____ x Fifteen = One Hundred Thirty-Five

14. _____ x Five = Sixty-Five

15. _____ x Three = Eighteen

16. _____ x Fifteen = Sixty

17. _____ x Nine = Ninety

18. _____ x Ten = Thirty

19. _____ x Six = Seventy-Two

20. _____ x Ten = Sixty

1. Fourteen x _____ = One Hundred Ninety-Six

2. Twelve x _____ = Seventy-Two

3. Ten x _____ = Seventy

4. Fifteen x _____ = Two Hundred Twenty-Five

5. Eleven x _____ = Ninety-Nine

6. Four x _____ = Sixteen

7. Seven x _____ = Seventy

8. Thirteen x _____ = Seventy-Eight

9. Eleven x _____ = Seventy-Seven

10. Fourteen x _____ = One Hundred Twenty-Six

11.　　　Fifteen x _____ = Forty-Five

12.　　　Three x _____ = Forty-Two

13.　　　Nine x _____ = One Hundred Thirty-Five

14.　　　Thirteen x _____ = Sixty-Five

15.　　　Six x _____ = Eighteen

16.　　　Four x _____ = Sixty

17.　　　Ten x _____ = Ninety

18.　　　Three x _____ = Thirty

19.　　　Twelve x _____ = Seventy-Two

20.　　　Six x _____ = Sixty

SKATEBOARDING IN DRESDEN, GERMANY

CHAPTER 3: K-3 MATH EXERCISES

DIVISIONS

1.   36 ÷ 18 = _____

2.   32 ÷ 8 = _____

3.   30 ÷ 3 = _____

4.   36 ÷ 9 = _____

5.   9 ÷ 3 = _____

6.   25 ÷ 5 = _____

7.   36 ÷ 3 = _____

8.   25 ÷ 5 = _____

9.   12 ÷ 6 = _____

10.   20 ÷ 5 = _____

11.   32 ÷ 4 = _____

12.   12 ÷ 6 = _____

13.   21 ÷ 3 = _____

14.   36 ÷ 9 = _____

15.   24 ÷ 8 = _____

16.	$27 \div 3 =$ _____
17.	$36 \div 18 =$ _____
18.	$30 \div 5 =$ _____
19.	$32 \div 16 =$ _____
20.	$24 \div 3 =$ _____
21.	$40 \div 10 =$ _____
22.	$12 \div 6 =$ _____
23.	$39 \div 13 =$ _____
24.	$14 \div 7 =$ _____
25.	$40 \div 5 =$ _____
26.	$32 \div 16 =$ _____
27.	$20 \div 4 =$ _____
28.	$24 \div 8 =$ _____
29.	$10 \div 5 =$ _____
30.	$32 \div 4 =$ _____
31.	$28 \div 7 =$ _____
32.	$33 \div 11 =$ _____

33.	$15 \div 5 =$ _____
34.	$12 \div 4 =$ _____
35.	$22 \div 11 =$ _____
36.	$39 \div 3 =$ _____
37.	$8 \div 4 =$ _____
38.	$38 \div 19 =$ _____
39.	$39 \div 3 =$ _____
40.	$32 \div 4 =$ _____
41.	$9 \div 3 =$ _____
42.	$40 \div 5 =$ _____
43.	$39 \div 13 =$ _____
44.	$12 \div 3 =$ _____
45.	$30 \div 5 =$ _____
46.	$12 \div 6 =$ _____
47.	$18 \div 9 =$ _____
48.	$32 \div 4 =$ _____
49.	$15 \div 5 =$ _____

50.	$21 \div 3 =$ _____
1.	$36 \div 18 = 2$
2.	$32 \div 8 = 4$
3.	$30 \div 3 = 10$
4.	$36 \div 9 = 4$
5.	$9 \div 3 = 3$
6.	$25 \div 5 = 5$
7.	$36 \div 3 = 12$
8.	$25 \div 5 = 5$
9.	$12 \div 6 = 2$
10.	$20 \div 5 = 4$
11.	$32 \div 4 = 8$
12.	$12 \div 6 = 2$
13.	$21 \div 3 = 7$
14.	$36 \div 9 = 4$
15.	$24 \div 8 = 3$
16.	$27 \div 3 = 9$

17.	$36 \div 18 = 2$
18.	$30 \div 5 = 6$
19.	$32 \div 16 = 2$
20.	$24 \div 3 = 8$
21.	$40 \div 10 = 4$
22.	$12 \div 6 = 2$
23.	$39 \div 13 = 3$
24.	$14 \div 7 = 2$
25.	$40 \div 5 = 8$
26.	$32 \div 16 = 2$
27.	$20 \div 4 = 5$
28.	$24 \div 8 = 3$
29.	$10 \div 5 = 2$
30.	$32 \div 4 = 8$
31.	$28 \div 7 = 4$
32.	$33 \div 11 = 3$
33.	$15 \div 5 = 3$

34.	$12 \div 4 = 3$
35.	$22 \div 11 = 2$
36.	$39 \div 3 = 13$
37.	$8 \div 4 = 2$
38.	$38 \div 19 = 2$
39.	$39 \div 3 = 13$
40.	$32 \div 4 = 8$
41.	$9 \div 3 = 3$
42.	$40 \div 5 = 8$
43.	$39 \div 13 = 3$
44.	$12 \div 3 = 4$
45.	$30 \div 5 = 6$
46.	$12 \div 6 = 2$
47.	$18 \div 9 = 2$
48.	$32 \div 4 = 8$
49.	$15 \div 5 = 3$
50.	$21 \div 3 = 7$

1.   ____ ÷ 18 = 2

2.   ____ ÷ 8 = 4

3.   ____ ÷ 3 = 10

4.   ____ ÷ 9 = 4

5.   ____ ÷ 3 = 3

6.   ____ ÷ 5 = 5

7.   ____ ÷ 3 = 12

8.   ____ ÷ 5 = 5

9.   ____ ÷ 6 = 2

10.   ____ ÷ 5 = 4

11.   ____ ÷ 4 = 8

12.   ____ ÷ 6 = 2

13.   ____ ÷ 3 = 7

14.   ____ ÷ 9 = 4

15.   ____ ÷ 8 = 3

16.   ____ ÷ 3 = 9

17.   ____ ÷ 18 = 2

18. _____ ÷ 5 = 6

19. _____ ÷ 16 = 2

20. _____ ÷ 3 = 8

21. _____ ÷ 10 = 4

22. _____ ÷ 6 = 2

23. _____ ÷ 13 = 3

24. _____ ÷ 7 = 2

25. _____ ÷ 5 = 8

26. _____ ÷ 16 = 2

27. _____ ÷ 4 = 5

28. _____ ÷ 8 = 3

29. _____ ÷ 5 = 2

30. _____ ÷ 4 = 8

31. _____ ÷ 7 = 4

32. _____ ÷ 11 = 3

33. _____ ÷ 5 = 3

34. _____ ÷ 4 = 3

35.   ____ $\div 11 = 2$

36.   ____ $\div 3 = 13$

37.   ____ $\div 4 = 2$

38.   ____ $\div 19 = 2$

39.   ____ $\div 3 = 13$

40.   ____ $\div 4 = 8$

41.   ____ $\div 3 = 3$

42.   ____ $\div 5 = 8$

43.   ____ $\div 13 = 3$

44.   ____ $\div 3 = 4$

45.   ____ $\div 5 = 6$

46.   ____ $\div 6 = 2$

47.   ____ $\div 9 = 2$

48.   ____ $\div 4 = 8$

49.   ____ $\div 5 = 3$

50.   ____ $\div 3 = 7$

1. $36 \div$ ____ $= 2$

2.	$32 \div \underline{\hspace{1cm}} = 4$
3.	$30 \div \underline{\hspace{1cm}} = 10$
4.	$36 \div \underline{\hspace{1cm}} = 4$
5.	$9 \div \underline{\hspace{1cm}} = 3$
6.	$25 \div \underline{\hspace{1cm}} = 5$
7.	$36 \div \underline{\hspace{1cm}} = 12$
8.	$25 \div \underline{\hspace{1cm}} = 5$
9.	$12 \div \underline{\hspace{1cm}} = 2$
10.	$20 \div \underline{\hspace{1cm}} = 4$
11.	$32 \div \underline{\hspace{1cm}} = 8$
12.	$12 \div \underline{\hspace{1cm}} = 2$
13.	$21 \div \underline{\hspace{1cm}} = 7$
14.	$36 \div \underline{\hspace{1cm}} = 4$
15.	$24 \div \underline{\hspace{1cm}} = 3$
16.	$27 \div \underline{\hspace{1cm}} = 9$
17.	$36 \div \underline{\hspace{1cm}} = 2$
18.	$30 \div \underline{\hspace{1cm}} = 6$

19.	$32 \div \underline{\quad} = 2$
20.	$24 \div \underline{\quad} = 8$
21.	$40 \div \underline{\quad} = 4$
22.	$12 \div \underline{\quad} = 2$
23.	$39 \div \underline{\quad} = 3$
24.	$14 \div \underline{\quad} = 2$
25.	$40 \div \underline{\quad} = 8$
26.	$32 \div \underline{\quad} = 2$
27.	$20 \div \underline{\quad} = 5$
28.	$24 \div \underline{\quad} = 3$
29.	$10 \div \underline{\quad} = 2$
30.	$32 \div \underline{\quad} = 8$
31.	$28 \div \underline{\quad} = 4$
32.	$33 \div \underline{\quad} = 3$
33.	$15 \div \underline{\quad} = 3$
34.	$12 \div \underline{\quad} = 3$
35.	$22 \div \underline{\quad} = 2$

36.	39 ÷ _____ = 13
37.	8 ÷ _____ = 2
38.	38 ÷ _____ = 2
39.	39 ÷ _____ = 13
40.	32 ÷ _____ = 8
41.	9 ÷ _____ = 3
42.	40 ÷ _____ = 8
43.	39 ÷ _____ = 3
44.	12 ÷ _____ = 4
45.	30 ÷ _____ = 6
46.	12 ÷ _____ = 2
47.	18 ÷ _____ = 2
48.	32 ÷ _____ = 8
49.	15 ÷ _____ = 3
50.	21 ÷ _____ = 7
1.	Thirty-Six ÷ Eighteen = _____
2.	Thirty-Two ÷ Eight = _____

3.	Thirty ÷ Three = _____
4.	Thirty-Six ÷ Nine = _____
5.	Nine ÷ Three = _____
6.	Twenty-Five ÷ Five = _____
7.	Thirty-Six ÷ Three = _____
8.	Twenty-Five ÷ Five = _____
9.	Twelve ÷ Six = _____
10.	Twenty ÷ Five = _____
11.	Thirty-Two ÷ Four = _____
12.	Twelve ÷ Six = _____
13.	Twenty-One ÷ Three = _____
14.	Thirty-Six ÷ Nine = _____
15.	Twenty-Four ÷ Eight = _____
16.	Twenty-Seven ÷ Three = _____
17.	Thirty-Six ÷ Eighteen = _____
18.	Thirty ÷ Five = _____
19.	Thirty-Two ÷ Sixteen = _____

20.	Twenty-Four ÷ Three = _____
1.	Thirty-Six ÷ Eighteen = Two
2.	Thirty-Two ÷ Eight = Four
3.	Thirty ÷ Three = Ten
4.	Thirty-Six ÷ Nine = Four
5.	Nine ÷ Three = Three
6.	Twenty-Five ÷ Five = Five
7.	Thirty-Six ÷ Three = Twelve
8.	Twenty-Five ÷ Five = Five
9.	Twelve ÷ Six = Two
10.	Twenty ÷ Five = Four
11.	Thirty-Two ÷ Four = Eight
12.	Twelve ÷ Six = Two
13.	Twenty-One ÷ Three = Seven
14.	Thirty-Six ÷ Nine = Four
15.	Twenty-Four ÷ Eight = Three
16.	Twenty-Seven ÷ Three = Nine

17.	Thirty-Six ÷ Eighteen = Two
18.	Thirty ÷ Five = Six
19.	Thirty-Two ÷ Sixteen = Two
20.	Twenty-Four ÷ Three = Eight
1.	_____ ÷ Eighteen = Two
2.	_____ ÷ Eight = Four
3.	_____ ÷ Three = Ten
4.	_____ ÷ Nine = Four
5.	_____ ÷ Three = Three
6.	_____ ÷ Five = Five
7.	_____ ÷ Three = Twelve
8.	_____ ÷ Five = Five
9.	_____ ÷ Six = Two
10.	_____ ÷ Five = Four
11.	_____ ÷ Four = Eight
12.	_____ ÷ Six = Two
13.	_____ ÷ Three = Seven

14.   _____ ÷ Nine = Four

15.   _____ ÷ Eight = Three

16.   _____ ÷ Three = Nine

17.   _____ ÷ Eighteen = Two

18.   _____ ÷ Five = Six

19.   _____ ÷ Sixteen = Two

20.   _____ ÷ Three = Eight

1.   Thirty-Six ÷ _____ = Two

2.   Thirty-Two ÷ _____ = Four

3.   Thirty ÷ _____ = Ten

4.   Thirty-Six ÷ _____ = Four

5.   Nine ÷ _____ = Three

6.   Twenty-Five ÷ _____ = Five

7.   Thirty-Six ÷ _____ = Twelve

8.   Twenty-Five ÷ _____ = Five

9.   Twelve ÷ _____ = Two

10.   Twenty ÷ _____ = Four

11.   Thirty-Two ÷ _____ = Eight

12.   Twelve ÷ _____ = Two

13.   Twenty-One ÷ _____ = Seven

14.   Thirty-Six ÷ _____ = Four

15.   Twenty-Four ÷ _____ = Three

16.   Twenty-Seven ÷ _____ = Nine

17.   Thirty-Six ÷ _____ = Two

18.   Thirty ÷ _____ = Six

19.   Thirty-Two ÷ _____ = Two

20.   Twenty-Four ÷ _____ = Eight

THE DEFUNCT THE SUN NEWSPAPER CLOCK IN NYC

CHAPTER 3: K-3 MATH EXERCISES

MIXED OPERATIONS

Computer Science operation symbols are used for multiplication (*) and division(/).

1.	20 - 13 + 7 = _____
2.	(18 + 17) - 15 = _____
3.	10 + 9 + 8 = _____
4.	(17 + 14) - 11 = _____
5.	20 + 18 - 6 = _____
6.	(12 * 7) + 5 = _____
7.	19 - (13 - 11) = _____
8.	(20 * 11) / 10 = _____
9.	(16 - 9) * 4 = _____
10.	(13 + 3) - 2 = _____
11.	(19 - 14) * 12 = _____
12.	13 - 6 / 2 = _____
13.	(20 + 17) - 15 = _____
14.	13 + (12 - 10) = _____

15.	(19 - 14) + 13 = _____
16.	20 + 18 - 17 = _____
17.	19 + 11 + 4 = _____
18.	(17 + 13) + 8 = _____
19.	(18 - 11) + 3 = _____
20.	20 - (13 - 11) = _____
21.	18 * 11 - 10 = _____
22.	20 * (14 - 9) = _____
23.	(19 + 16) - 10 = _____
24.	(11 - 10) + 6 = _____
25.	(18 - 12) * 9 = _____
26.	17 * 7 + 4 = _____
27.	14 * 11 - 9 = _____
28.	20 + 16 - 2 = _____
29.	15 + 10 - 8 = _____
30.	20 - 13 + 9 = _____
31.	19 + (7 - 6) = _____

32.   13 * 12 - 11 = _____

33.   (17 + 11) + 8 = _____

34.   13 + (8 - 5) = _____

35.   20 - (17 - 10) = _____

36.   (12 - 10) * 7 = _____

37.   19 - 18 + 17 = _____

38.   16 * 8 - 4 = _____

39.   (20 + 12) + 6 = _____

40.   (14 / 7) + 4 = _____

41.   17 + 15 * 7 = _____

42.   (15 - 7) + 6 = _____

43.   (17 - 10) * 5 = _____

44.   (20 + 18) - 10 = _____

45.   (13 - 11) * 9 = _____

46.   20 + (18 + 15) = _____

47.   15 * 5 * 2 = _____

48.   16 + 8 - 6 = _____

49.	**(15 + 10) + 8 = ____**
50.	**8 * 5 * 4 = ____**
51.	**12 - (10 - 8) = ____**
52.	**(16 - 14) + 11 = ____**
53.	**17 - 15 + 8 = ____**
54.	**20 * (14 - 12) = ____**
55.	**19 * (16 - 9) = ____**
56.	**(9 * 6) - 4 = ____**
57.	**(19 + 16) - 15 = ____**
58.	**18 + 17 * 8 = ____**
59.	**(19 + 15) + 9 = ____**
60.	**(20 + 19) - 17 = ____**
61.	**(15 + 13) - 7 = ____**
62.	**(6 - 4) + 3 = ____**
63.	**20 * (15 - 13) = ____**
64.	**14 + 12 * 8 = ____**
65.	**18 - 9 + 4 = ____**

66.	13 - 10 + 9 = _____
67.	10 * 9 + 4 = _____
68.	17 - (15 - 12) = _____
69.	(19 - 13) + 11 = _____
70.	12 / 6 * 5 = _____
71.	(16 * 12) - 9 = _____
72.	12 + 8 + 6 = _____
73.	16 * 11 - 3 = _____
74.	(11 - 6) + 5 = _____
75.	(14 * 12) + 11 = _____
76.	(13 + 11) + 10 = _____
77.	19 + 10 - 7 = _____
78.	20 + (6 * 3) = _____
79.	(18 - 13) * 8 = _____
80.	15 + 9 + 3 = _____
81.	16 + 11 - 10 = _____
82.	13 + 9 - 8 = _____

83.	$(10 + 6) * 2 = \rule{2cm}{0.4pt}$
84.	$19 + 13 * 12 = \rule{2cm}{0.4pt}$
85.	$(20 - 16) + 6 = \rule{2cm}{0.4pt}$
86.	$9 + (5 * 4) = \rule{2cm}{0.4pt}$
87.	$13 - 10 + 8 = \rule{2cm}{0.4pt}$
88.	$12 + 11 * 6 = \rule{2cm}{0.4pt}$
89.	$18 + 15 + 13 = \rule{2cm}{0.4pt}$
90.	$(20 - 12) * 6 = \rule{2cm}{0.4pt}$
91.	$(17 + 11) + 8 = \rule{2cm}{0.4pt}$
92.	$19 - 5 + 3 = \rule{2cm}{0.4pt}$
93.	$16 + (14 + 13) = \rule{2cm}{0.4pt}$
94.	$18 + 11 * 7 = \rule{2cm}{0.4pt}$
95.	$17 + (10 - 3) = \rule{2cm}{0.4pt}$
96.	$(13 * 12) - 8 = \rule{2cm}{0.4pt}$
97.	$19 + 15 - 14 = \rule{2cm}{0.4pt}$
98.	$(15 * 9) + 4 = \rule{2cm}{0.4pt}$
99.	$14 / (13 - 11) = \rule{2cm}{0.4pt}$

100.	19 * (8 - 2) = _____
1.	20 - 13 + 7 = 14
2.	(18 + 17) - 15 = 20
3.	10 + 9 + 8 = 27
4.	(17 + 14) - 11 = 20
5.	20 + 18 - 6 = 32
6.	(12 * 7) + 5 = 89
7.	19 - (13 - 11) = 17
8.	(20 * 11) / 10 = 22
9.	(16 - 9) * 4 = 28
10.	(13 + 3) - 2 = 14
11.	(19 - 14) * 12 = 60
12.	13 - 6 / 2 = 10
13.	(20 + 17) - 15 = 22
14.	13 + (12 - 10) = 15
15.	(19 - 14) + 13 = 18
16.	20 + 18 - 17 = 21

17.	$19 + 11 + 4 = 34$
18.	$(17 + 13) + 8 = 38$
19.	$(18 - 11) + 3 = 10$
20.	$20 - (13 - 11) = 18$
21.	$18 * 11 - 10 = 188$
22.	$20 * (14 - 9) = 100$
23.	$(19 + 16) - 10 = 25$
24.	$(11 - 10) + 6 = 7$
25.	$(18 - 12) * 9 = 54$
26.	$17 * 7 + 4 = 123$
27.	$14 * 11 - 9 = 145$
28.	$20 + 16 - 2 = 34$
29.	$15 + 10 - 8 = 17$
30.	$20 - 13 + 9 = 16$
31.	$19 + (7 - 6) = 20$
32.	$13 * 12 - 11 = 145$
33.	$(17 + 11) + 8 = 36$

34.	13 + (8 - 5) = 16
35.	20 - (17 - 10) = 13
36.	(12 - 10) * 7 = 14
37.	19 - 18 + 17 = 18
38.	16 * 8 - 4 = 124
39.	(20 + 12) + 6 = 38
40.	(14 / 7) + 4 = 6
41.	17 + 15 * 7 = 122
42.	(15 - 7) + 6 = 14
43.	(17 - 10) * 5 = 35
44.	(20 + 18) - 10 = 28
45.	(13 - 11) * 9 = 18
46.	20 + (18 + 15) = 53
47.	15 * 5 * 2 = 150
48.	16 + 8 - 6 = 18
49.	(15 + 10) + 8 = 33
50.	8 * 5 * 4 = 160

51.	12 - (10 - 8) = 10
52.	(16 - 14) + 11 = 13
53.	17 - 15 + 8 = 10
54.	20 * (14 - 12) = 40
55.	19 * (16 - 9) = 133
56.	(9 * 6) - 4 = 50
57.	(19 + 16) - 15 = 20
58.	18 + 17 * 8 = 154
59.	(19 + 15) + 9 = 43
60.	(20 + 19) - 17 = 22
61.	(15 + 13) - 7 = 21
62.	(6 - 4) + 3 = 5
63.	20 * (15 - 13) = 40
64.	14 + 12 * 8 = 110
65.	18 - 9 + 4 = 13
66.	13 - 10 + 9 = 12
67.	10 * 9 + 4 = 94

68.	$17 - (15 - 12) = 14$
69.	$(19 - 13) + 11 = 17$
70.	$12 / 6 * 5 = 10$
71.	$(16 * 12) - 9 = 183$
72.	$12 + 8 + 6 = 26$
73.	$16 * 11 - 3 = 173$
74.	$(11 - 6) + 5 = 10$
75.	$(14 * 12) + 11 = 179$
76.	$(13 + 11) + 10 = 34$
77.	$19 + 10 - 7 = 22$
78.	$20 + (6 * 3) = 38$
79.	$(18 - 13) * 8 = 40$
80.	$15 + 9 + 3 = 27$
81.	$16 + 11 - 10 = 17$
82.	$13 + 9 - 8 = 14$
83.	$(10 + 6) * 2 = 32$
84.	$19 + 13 * 12 = 175$

85.	(20 - 16) + 6 = 10
86.	9 + (5 * 4) = 29
87.	13 - 10 + 8 = 11
88.	12 + 11 * 6 = 78
89.	18 + 15 + 13 = 46
90.	(20 - 12) * 6 = 48
91.	(17 + 11) + 8 = 36
92.	19 - 5 + 3 = 17
93.	16 + (14 + 13) = 43
94.	18 + 11 * 7 = 95
95.	17 + (10 - 3) = 24
96.	(13 * 12) - 8 = 148
97.	19 + 15 - 14 = 20
98.	(15 * 9) + 4 = 139
99.	14 / (13 - 11) = 7
100.	19 * (8 - 2) = 114
1.	20 - 13 + _____ = 14

2. $(18 + \underline{\hspace{1cm}}) - 15 = 20$

3. $10 + 9 + \underline{\hspace{1cm}} = 27$

4. $(17 + \underline{\hspace{1cm}}) - 11 = 20$

5. $\underline{\hspace{1cm}} + 18 - 6 = 32$

6. $(12 * 7) + \underline{\hspace{1cm}} = 89$

7. $19 - (\underline{\hspace{1cm}} - 11) = 17$

8. $(20 * \underline{\hspace{1cm}}) / 10 = 22$

9. $(16 - 9) * \underline{\hspace{1cm}} = 28$

10. $(\underline{\hspace{1cm}} + 3) - 2 = 14$

11. $(19 - 14) * \underline{\hspace{1cm}} = 60$

12. $13 - \underline{\hspace{1cm}} / 2 = 10$

13. $(20 + \underline{\hspace{1cm}}) - 15 = 22$

14. $13 + (12 - \underline{\hspace{1cm}}) = 15$

15. $(19 - \underline{\hspace{1cm}}) + 13 = 18$

16. $\underline{\hspace{1cm}} + 18 - 17 = 21$

17. $19 + 11 + \underline{\hspace{1cm}} = 34$

18. $(\underline{\hspace{1cm}} + 13) + 8 = 38$

19.	(_____ - 11) + 3 = 10
20.	20 - (_____ - 11) = 18
21.	18 * 11 - _____ = 188
22.	20 * (_____ - 9) = 100
23.	(_____ + 16) - 10 = 25
24.	(11 - 10) + _____ = 7
25.	(18 - _____) * 9 = 54
26.	17 * _____ + 4 = 123
27.	14 * 11 - _____ = 145
28.	20 + _____ - 2 = 34
29.	_____ + 10 - 8 = 17
30.	_____ - 13 + 9 = 16
31.	_____ + (7 - 6) = 20
32.	13 * _____ - 11 = 145
33.	(17 + 11) + _____ = 36
34.	13 + (_____ - 5) = 16
35.	20 - (17 - _____) = 13

36.	(_____ - 10) * 7 = 14
37.	_____ - 18 + 17 = 18
38.	16 * _____ - 4 = 124
39.	(20 + 12) + _____ = 38
40.	(14 / _____) + 4 = 6
41.	17 + 15 * _____ = 122
42.	(15 - _____) + 6 = 14
43.	(17 - _____) * 5 = 35
44.	(20 + _____) - 10 = 28
45.	(13 - 11) * _____ = 18
46.	20 + (_____ + 15) = 53
47.	15 * 5 * _____ = 150
48.	_____ + 8 - 6 = 18
49.	(_____ + 10) + 8 = 33
50.	8 * _____ * 4 = 160
51.	12 - (_____ - 8) = 10
52.	(_____ - 14) + 11 = 13

53. _____ - 15 + 8 = 10

54. 20 * (14 - _____) = 40

55. 19 * (16 - _____) = 133

56. (9 * 6) - _____ = 50

57. (19 + 16) - _____ = 20

58. 18 + _____ * 8 = 154

59. (_____ + 15) + 9 = 43

60. (_____ + 19) - 17 = 22

61. (_____ + 13) - 7 = 21

62. (6 - _____) + 3 = 5

63. 20 * (15 - _____) = 40

64. 14 + 12 * _____ = 110

65. 18 - _____ + 4 = 13

66. 13 - 10 + _____ = 12

67. 10 * _____ + 4 = 94

68. 17 - (_____ - 12) = 14

69. (19 - _____) + 11 = 17

70.	_____ / 6 * 5 = 10
71.	(_____ * 12) - 9 = 183
72.	12 + _____ + 6 = 26
73.	_____ * 11 - 3 = 173
74.	(11 - _____) + 5 = 10
75.	(_____ * 12) + 11 = 179
76.	(_____ + 11) + 10 = 34
77.	_____ + 10 - 7 = 22
78.	20 + (_____ * 3) = 38
79.	(_____ - 13) * 8 = 40
80.	15 + 9 + _____ = 27
81.	16 + 11 - _____ = 17
82.	_____ + 9 - 8 = 14
83.	(10 + _____) * 2 = 32
84.	_____ + 13 * 12 = 175
85.	(20 - _____) + 6 = 10
86.	9 + (_____ * 4) = 29

87.	13 - 10 + _____ = 11
88.	_____ + 11 * 6 = 78
89.	_____ + 15 + 13 = 46
90.	(20 - 12) * _____ = 48
91.	(17 + _____) + 8 = 36
92.	19 - _____ + 3 = 17
93.	16 + (_____ + 13) = 43
94.	18 + _____ * 7 = 95
95.	_____ + (10 - 3) = 24
96.	(13 * 12) - _____ = 148
97.	19 + 15 - _____ = 20
98.	(15 * 9) + _____ = 139
99.	_____ / (13 - 11) = 7
100.	19 * (_____ - 2) = 114

ELEPHANT HOLDING THE BIG APPLE (NICK FOR NYC) ON COLUMBUS CIRCLE

CHAPTER 3: K-3 MATH EXERCISES

CORPORATE JET AT TETERBORO AIRPORT, NEW JERSEY

CHAPTER 3: K-3 MATH EXERCISES

CHAPTER 4: K-4 MATH EXERCISES

ADDITIONS

1.   259 + 117 = _____

2.   127 + 102 = _____

3.   125 + 54 = _____

4.   213 + 12 = _____

5.   279 + 221 = _____

6.   283 + 282 = _____

7.   85 + 291 = _____

8.   290 + 114 = _____

9.   234 + 179 = _____

10.   176 + 231 = _____

11.   31 + 144 = _____

12.   282 + 65 = _____

13.   41 + 173 = _____

14.   53 + 189 = _____

15. 156 + 250 = ____

16. 179 + 247 = ____

17. 21 + 159 = ____

18. 212 + 95 = ____

19. 231 + 126 = ____

20. 43 + 43 = ____

21. 265 + 163 = ____

22. 146 + 84 = ____

23. 36 + 219 = ____

24. 157 + 91 = ____

25. 196 + 113 = ____

26. 172 + 25 = ____

27. 216 + 136 = ____

28. 134 + 29 = ____

29. 204 + 197 = ____

30. 24 + 178 = ____

31. 64 + 198 = ____

32.   130 + 132 = _____

33.   272 + 130 = _____

34.   192 + 216 = _____

35.   57 + 212 = _____

36.   263 + 129 = _____

37.   81 + 247 = _____

38.   192 + 286 = _____

39.   236 + 116 = _____

40.   213 + 195 = _____

41.   251 + 286 = _____

42.   132 + 156 = _____

43.   105 + 207 = _____

44.   159 + 71 = _____

45.   259 + 91 = _____

46.   207 + 185 = _____

47.   133 + 241 = _____

48.   247 + 198 = _____

49.	232 + 299 = _____
50.	102 + 9 = _____
1.	259 + 117 = 376
2.	127 + 102 = 229
3.	125 + 54 = 179
4.	213 + 12 = 225
5.	279 + 221 = 500
6.	283 + 282 = 565
7.	85 + 291 = 376
8.	290 + 114 = 404
9.	234 + 179 = 413
10.	176 + 231 = 407
11.	31 + 144 = 175
12.	282 + 65 = 347
13.	41 + 173 = 214
14.	53 + 189 = 242
15.	156 + 250 = 406

16.	179 + 247 = 426
17.	21 + 159 = 180
18.	212 + 95 = 307
19.	231 + 126 = 357
20.	43 + 43 = 86
21.	265 + 163 = 428
22.	146 + 84 = 230
23.	36 + 219 = 255
24.	157 + 91 = 248
25.	196 + 113 = 309
26.	172 + 25 = 197
27.	216 + 136 = 352
28.	134 + 29 = 163
29.	204 + 197 = 401
30.	24 + 178 = 202
31.	64 + 198 = 262
32.	130 + 132 = 262

33.	272 + 130 = 402
34.	192 + 216 = 408
35.	57 + 212 = 269
36.	263 + 129 = 392
37.	81 + 247 = 328
38.	192 + 286 = 478
39.	236 + 116 = 352
40.	213 + 195 = 408
41.	251 + 286 = 537
42.	132 + 156 = 288
43.	105 + 207 = 312
44.	159 + 71 = 230
45.	259 + 91 = 350
46.	207 + 185 = 392
47.	133 + 241 = 374
48.	247 + 198 = 445
49.	232 + 299 = 531

50.   102 + 9 = 111

1.   _____ + 117 = 376

2.   _____ + 102 = 229

3.   _____ + 54 = 179

4.   _____ + 12 = 225

5.   _____ + 221 = 500

6.   _____ + 282 = 565

7.   _____ + 291 = 376

8.   _____ + 114 = 404

9.   _____ + 179 = 413

10.   _____ + 231 = 407

11.   _____ + 144 = 175

12.   _____ + 65 = 347

13.   _____ + 173 = 214

14.   _____ + 189 = 242

15.   _____ + 250 = 406

16.   _____ + 247 = 426

17. ____ + 159 = 180

18. ____ + 95 = 307

19. ____ + 126 = 357

20. ____ + 43 = 86

21. ____ + 163 = 428

22. ____ + 84 = 230

23. ____ + 219 = 255

24. ____ + 91 = 248

25. ____ + 113 = 309

26. ____ + 25 = 197

27. ____ + 136 = 352

28. ____ + 29 = 163

29. ____ + 197 = 401

30. ____ + 178 = 202

31. ____ + 198 = 262

32. ____ + 132 = 262

33. ____ + 130 = 402

34. _____ + 216 = 408

35. _____ + 212 = 269

36. _____ + 129 = 392

37. _____ + 247 = 328

38. _____ + 286 = 478

39. _____ + 116 = 352

40. _____ + 195 = 408

41. _____ + 286 = 537

42. _____ + 156 = 288

43. _____ + 207 = 312

44. _____ + 71 = 230

45. _____ + 91 = 350

46. _____ + 185 = 392

47. _____ + 241 = 374

48. _____ + 198 = 445

49. _____ + 299 = 531

50. _____ + 9 = 111

1. 259 + _____ = 376

2. 127 + _____ = 229

3. 125 + _____ = 179

4. 213 + _____ = 225

5. 279 + _____ = 500

6. 283 + _____ = 565

7. 85 + _____ = 376

8. 290 + _____ = 404

9. 234 + _____ = 413

10. 176 + _____ = 407

11. 31 + _____ = 175

12. 282 + _____ = 347

13. 41 + _____ = 214

14. 53 + _____ = 242

15. 156 + _____ = 406

16. 179 + _____ = 426

17. 21 + _____ = 180

18. $212 + \_\_\_\_ = 307$

19. $231 + \_\_\_\_ = 357$

20. $43 + \_\_\_\_ = 86$

21. $265 + \_\_\_\_ = 428$

22. $146 + \_\_\_\_ = 230$

23. $36 + \_\_\_\_ = 255$

24. $157 + \_\_\_\_ = 248$

25. $196 + \_\_\_\_ = 309$

26. $172 + \_\_\_\_ = 197$

27. $216 + \_\_\_\_ = 352$

28. $134 + \_\_\_\_ = 163$

29. $204 + \_\_\_\_ = 401$

30. $24 + \_\_\_\_ = 202$

31. $64 + \_\_\_\_ = 262$

32. $130 + \_\_\_\_ = 262$

33. $272 + \_\_\_\_ = 402$

34. $192 + \_\_\_\_ = 408$

35.	57 + ____ = 269
36.	263 + ____ = 392
37.	81 + ____ = 328
38.	192 + ____ = 478
39.	236 + ____ = 352
40.	213 + ____ = 408
41.	251 + ____ = 537
42.	132 + ____ = 288
43.	105 + ____ = 312
44.	159 + ____ = 230
45.	259 + ____ = 350
46.	207 + ____ = 392
47.	133 + ____ = 374
48.	247 + ____ = 445
49.	232 + ____ = 531
50.	102 + ____ = 111
1.	Two Hundred Fifty-Nine + One Hundred Seventeen = ____

2.   One Hundred Twenty-Seven + One Hundred Two = _____

3.   One Hundred Twenty-Five + Fifty-Four = _____

4.   Two Hundred Thirteen + Twelve = _____

5.   Two Hundred Seventy-Nine + Two Hundred Twenty-One = _____

6.   Two Hundred Eighty-Three + Two Hundred Eighty-Two = _____

7.   Eighty-Five + Two Hundred Ninety-One = _____

8.   Two Hundred Ninety + One Hundred Fourteen = _____

9.   Two Hundred Thirty-Four + One Hundred Seventy-Nine = _____

10.   One Hundred Seventy-Six + Two Hundred Thirty-One = _____

11.   Thirty-One + One Hundred Forty-Four = _____

12.   Two Hundred Eighty-Two + Sixty-Five = _____

13.   Forty-One + One Hundred Seventy-Three = _____

14.   Fifty-Three + One Hundred Eighty-Nine = _____

15.   One Hundred Fifty-Six + Two Hundred Fifty = _____

16. One Hundred Seventy-Nine + Two Hundred Forty-Seven = _____

17. Twenty-One + One Hundred Fifty-Nine = _____

18. Two Hundred Twelve + Ninety-Five = _____

19. Two Hundred Thirty-One + One Hundred Twenty-Six = _____

20. Forty-Three + Forty-Three = _____

21. Two Hundred Sixty-Five + One Hundred Sixty-Three = _____

22. One Hundred Forty-Six + Eighty-Four = _____

23. Thirty-Six + Two Hundred Nineteen = _____

24. One Hundred Fifty-Seven + Ninety-One = _____

25. One Hundred Ninety-Six + One Hundred Thirteen = _____

26. One Hundred Seventy-Two + Twenty-Five = _____

27. Two Hundred Sixteen + One Hundred Thirty-Six = _____

28. One Hundred Thirty-Four + Twenty-Nine = _____

29. Two Hundred Four + One Hundred Ninety-Seven = _____

30.   Twenty-Four + One Hundred Seventy-Eight = _____

1.   Two Hundred Fifty-Nine + One Hundred Seventeen = Three Hundred Seventy-Six

2.   One Hundred Twenty-Seven + One Hundred Two = Two Hundred Twenty-Nine

3.   One Hundred Twenty-Five + Fifty-Four = One Hundred Seventy-Nine

4.   Two Hundred Thirteen + Twelve = Two Hundred Twenty-Five

5.   Two Hundred Seventy-Nine + Two Hundred Twenty-One = Five Hundred

6.   Two Hundred Eighty-Three + Two Hundred Eighty-Two = Five Hundred Sixty-Five

7.   Eighty-Five + Two Hundred Ninety-One = Three Hundred Seventy-Six

8.   Two Hundred Ninety + One Hundred Fourteen = Four Hundred Four

9.   Two Hundred Thirty-Four + One Hundred Seventy-Nine = Four Hundred Thirteen

10.   One Hundred Seventy-Six + Two Hundred Thirty-One = Four Hundred Seven

11.	Thirty-One + One Hundred Forty-Four = One Hundred Seventy-Five
12.	Two Hundred Eighty-Two + Sixty-Five = Three Hundred Forty-Seven
13.	Forty-One + One Hundred Seventy-Three = Two Hundred Fourteen
14.	Fifty-Three + One Hundred Eighty-Nine = Two Hundred Forty-Two
15.	One Hundred Fifty-Six + Two Hundred Fifty = Four Hundred Six
16.	One Hundred Seventy-Nine + Two Hundred Forty-Seven = Four Hundred Twenty-Six
17.	Twenty-One + One Hundred Fifty-Nine = One Hundred Eighty
18.	Two Hundred Twelve + Ninety-Five = Three Hundred Seven
19.	Two Hundred Thirty-One + One Hundred Twenty-Six = Three Hundred Fifty-Seven
20.	Forty-Three + Forty-Three = Eighty-Six
21.	Two Hundred Sixty-Five + One Hundred Sixty-Three = Four Hundred Twenty-Eight
22.	One Hundred Forty-Six + Eighty-Four = Two Hundred Thirty

23.	Thirty-Six + Two Hundred Nineteen = Two Hundred Fifty-Five
24.	One Hundred Fifty-Seven + Ninety-One = Two Hundred Forty-Eight
25.	One Hundred Ninety-Six + One Hundred Thirteen = Three Hundred Nine
26.	One Hundred Seventy-Two + Twenty-Five = One Hundred Ninety-Seven
27.	Two Hundred Sixteen + One Hundred Thirty-Six = Three Hundred Fifty-Two
28.	One Hundred Thirty-Four + Twenty-Nine = One Hundred Sixty-Three
29.	Two Hundred Four + One Hundred Ninety-Seven = Four Hundred One
30.	Twenty-Four + One Hundred Seventy-Eight = Two Hundred Two
1.	_____ + One Hundred Seventeen = Three Hundred Seventy-Six
2.	_____ + One Hundred Two = Two Hundred Twenty-Nine
3.	_____ + Fifty-Four = One Hundred Seventy-Nine
4.	_____ + Twelve = Two Hundred Twenty-Five
5.	_____ + Two Hundred Twenty-One = Five Hundred

6.   _____ + Two Hundred Eighty-Two = Five Hundred Sixty-Five

7.   _____ + Two Hundred Ninety-One = Three Hundred Seventy-Six

8.   _____ + One Hundred Fourteen = Four Hundred Four

9.   _____ + One Hundred Seventy-Nine = Four Hundred Thirteen

10.   _____ + Two Hundred Thirty-One = Four Hundred Seven

11.   _____ + One Hundred Forty-Four = One Hundred Seventy-Five

12.   _____ + Sixty-Five = Three Hundred Forty-Seven

13.   _____ + One Hundred Seventy-Three = Two Hundred Fourteen

14.   _____ + One Hundred Eighty-Nine = Two Hundred Forty-Two

15.   _____ + Two Hundred Fifty = Four Hundred Six

16.   _____ + Two Hundred Forty-Seven = Four Hundred Twenty-Six

17.   _____ + One Hundred Fifty-Nine = One Hundred Eighty

18.   _____ + Ninety-Five = Three Hundred Seven

19.   _____ + One Hundred Twenty-Six = Three Hundred Fifty-

	Seven
20.	_____ + Forty-Three = Eighty-Six
21.	_____ + One Hundred Sixty-Three = Four Hundred Twenty-Eight
22.	_____ + Eighty-Four = Two Hundred Thirty
23.	_____ + Two Hundred Nineteen = Two Hundred Fifty-Five
24.	_____ + Ninety-One = Two Hundred Forty-Eight
25.	_____ + One Hundred Thirteen = Three Hundred Nine
26.	_____ + Twenty-Five = One Hundred Ninety-Seven
27.	_____ + One Hundred Thirty-Six = Three Hundred Fifty-Two
28.	_____ + Twenty-Nine = One Hundred Sixty-Three
29.	_____ + One Hundred Ninety-Seven = Four Hundred One
30.	_____ + One Hundred Seventy-Eight = Two Hundred Two
1.	Two Hundred Fifty-Nine + _____ = Three Hundred Seventy-Six
2.	One Hundred Twenty-Seven + _____ = Two Hundred Twenty-Nine
3.	One Hundred Twenty-Five + _____ = One Hundred Seventy-Nine
4.	Two Hundred Thirteen + _____ = Two Hundred Twenty-

	Five
5.	Two Hundred Seventy-Nine + ____ = Five Hundred
6.	Two Hundred Eighty-Three + ____ = Five Hundred Sixty-Five
7.	Eighty-Five + ____ = Three Hundred Seventy-Six
8.	Two Hundred Ninety + ____ = Four Hundred Four
9.	Two Hundred Thirty-Four + ____ = Four Hundred Thirteen
10.	One Hundred Seventy-Six + ____ = Four Hundred Seven
11.	Thirty-One + ____ = One Hundred Seventy-Five
12.	Two Hundred Eighty-Two + ____ = Three Hundred Forty-Seven
13.	Forty-One + ____ = Two Hundred Fourteen
14.	Fifty-Three + ____ = Two Hundred Forty-Two
15.	One Hundred Fifty-Six + ____ = Four Hundred Six
16.	One Hundred Seventy-Nine + ____ = Four Hundred Twenty-Six
17.	Twenty-One + ____ = One Hundred Eighty
18.	Two Hundred Twelve + ____ = Three Hundred Seven
19.	Two Hundred Thirty-One + ____ = Three Hundred Fifty-Seven

20.	Forty-Three + _____ = Eighty-Six
21.	Two Hundred Sixty-Five + _____ = Four Hundred Twenty-Eight
22.	One Hundred Forty-Six + _____ = Two Hundred Thirty
23.	Thirty-Six + _____ = Two Hundred Fifty-Five
24.	One Hundred Fifty-Seven + _____ = Two Hundred Forty-Eight
25.	One Hundred Ninety-Six + _____ = Three Hundred Nine
26.	One Hundred Seventy-Two + _____ = One Hundred Ninety-Seven
27.	Two Hundred Sixteen + _____ = Three Hundred Fifty-Two
28.	One Hundred Thirty-Four + _____ = One Hundred Sixty-Three
29.	Two Hundred Four + _____ = Four Hundred One
30.	Twenty-Four + _____ = Two Hundred Two

EMPIRE STATE & CHRYSLER BUILDINGS BUILT AROUND 1935 IN NEW YORK CITY

CHAPTER 4: K-4 MATH EXERCISES

SUBTRACTIONS

1.   193 - 15 = _____

2.   238 - 125 = _____

3.   282 - 218 = _____

4.   251 - 39 = _____

5.   171 - 144 = _____

6.   244 - 58 = _____

7.   214 - 12 = _____

8.   78 - 26 = _____

9.   242 - 108 = _____

10.   213 - 92 = _____

11.   237 - 102 = _____

12.   221 - 28 = _____

13.   285 - 276 = _____

14.   263 - 257 = _____

15.   202 - 21 = _____

16. 262 - 215 = _____

17. 111 - 53 = _____

18. 126 - 79 = _____

19. 112 - 34 = _____

20. 292 - 189 = _____

21. 127 - 10 = _____

22. 168 - 46 = _____

23. 270 - 156 = _____

24. 169 - 6 = _____

25. 199 - 140 = _____

26. 169 - 95 = _____

27. 283 - 242 = _____

28. 232 - 66 = _____

29. 194 - 147 = _____

30. 185 - 164 = _____

31. 286 - 247 = _____

32. 210 - 199 = _____

33. 200 - 30 = _____

34.   **175 - 172 = ____**

35.   **281 - 170 = ____**

36.   **93 - 83 = ____**

37.   **295 - 261 = ____**

38.   **122 - 32 = ____**

39.   **298 - 130 = ____**

40.   **277 - 66 = ____**

41.   **260 - 80 = ____**

42.   **70 - 7 = ____**

43.   **268 - 39 = ____**

44.   **207 - 98 = ____**

45.   **217 - 84 = ____**

46.   **81 - 55 = ____**

47.   **256 - 182 = ____**

48.   **225 - 43 = ____**

49.   **78 - 38 = ____**

50.   **246 - 44 = ____**

1.   **193 - 15 = 178**

2.	238 - 125 = 113
3.	282 - 218 = 64
4.	251 - 39 = 212
5.	171 - 144 = 27
6.	244 - 58 = 186
7.	214 - 12 = 202
8.	78 - 26 = 52
9.	242 - 108 = 134
10.	213 - 92 = 121
11.	237 - 102 = 135
12.	221 - 28 = 193
13.	285 - 276 = 9
14.	263 - 257 = 6
15.	202 - 21 = 181
16.	262 - 215 = 47
17.	111 - 53 = 58
18.	126 - 79 = 47
19.	112 - 34 = 78

20.	292 - 189 = 103
21.	127 - 10 = 117
22.	168 - 46 = 122
23.	270 - 156 = 114
24.	169 - 6 = 163
25.	199 - 140 = 59
26.	169 - 95 = 74
27.	283 - 242 = 41
28.	232 - 66 = 166
29.	194 - 147 = 47
30.	185 - 164 = 21
31.	286 - 247 = 39
32.	210 - 199 = 11
33.	200 - 30 = 170
34.	175 - 172 = 3
35.	281 - 170 = 111
36.	93 - 83 = 10
37.	295 - 261 = 34

38.	122 - 32 = 90
39.	298 - 130 = 168
40.	277 - 66 = 211
41.	260 - 80 = 180
42.	70 - 7 = 63
43.	268 - 39 = 229
44.	207 - 98 = 109
45.	217 - 84 = 133
46.	81 - 55 = 26
47.	256 - 182 = 74
48.	225 - 43 = 182
49.	78 - 38 = 40
50.	246 - 44 = 202
1.	_____ - 15 = 178
2.	_____ - 125 = 113
3.	_____ - 218 = 64
4.	_____ - 39 = 212
5.	_____ - 144 = 27

6.   _____ - 58 = 186

7.   _____ - 12 = 202

8.   _____ - 26 = 52

9.   _____ - 108 = 134

10.   _____ - 92 = 121

11.   _____ - 102 = 135

12.   _____ - 28 = 193

13.   _____ - 276 = 9

14.   _____ - 257 = 6

15.   _____ - 21 = 181

16.   _____ - 215 = 47

17.   _____ - 53 = 58

18.   _____ - 79 = 47

19.   _____ - 34 = 78

20.   _____ - 189 = 103

21.   _____ - 10 = 117

22.   _____ - 46 = 122

23.   _____ - 156 = 114

24. _____ - 6 = 163

25. _____ - 140 = 59

26. _____ - 95 = 74

27. _____ - 242 = 41

28. _____ - 66 = 166

29. _____ - 147 = 47

30. _____ - 164 = 21

31. _____ - 247 = 39

32. _____ - 199 = 11

33. _____ - 30 = 170

34. _____ - 172 = 3

35. _____ - 170 = 111

36. _____ - 83 = 10

37. _____ - 261 = 34

38. _____ - 32 = 90

39. _____ - 130 = 168

40. _____ - 66 = 211

41. _____ - 80 = 180

42.   _____ - 7 = 63

43.   _____ - 39 = 229

44.   _____ - 98 = 109

45.   _____ - 84 = 133

46.   _____ - 55 = 26

47.   _____ - 182 = 74

48.   _____ - 43 = 182

49.   _____ - 38 = 40

50.   _____ - 44 = 202

1.   193 - _____ = 178

2.   238 - _____ = 113

3.   282 - _____ = 64

4.   251 - _____ = 212

5.   171 - _____ = 27

6.   244 - _____ = 186

7.   214 - _____ = 202

8.   78 - _____ = 52

9.   242 - _____ = 134

10. 213 - _____ = 121

11. 237 - _____ = 135

12. 221 - _____ = 193

13. 285 - _____ = 9

14. 263 - _____ = 6

15. 202 - _____ = 181

16. 262 - _____ = 47

17. 111 - _____ = 58

18. 126 - _____ = 47

19. 112 - _____ = 78

20. 292 - _____ = 103

21. 127 - _____ = 117

22. 168 - _____ = 122

23. 270 - _____ = 114

24. 169 - _____ = 163

25. 199 - _____ = 59

26. 169 - _____ = 74

27. 283 - _____ = 41

28.   232 - _____ = 166

29.   194 - _____ = 47

30.   185 - _____ = 21

31.   286 - _____ = 39

32.   210 - _____ = 11

33.   200 - _____ = 170

34.   175 - _____ = 3

35.   281 - _____ = 111

36.   93 - _____ = 10

37.   295 - _____ = 34

38.   122 - _____ = 90

39.   298 - _____ = 168

40.   277 - _____ = 211

41.   260 - _____ = 180

42.   70 - _____ = 63

43.   268 - _____ = 229

44.   207 - _____ = 109

45.   217 - _____ = 133

46.   81 - ____ = 26

47.   256 - ____ = 74

48.   225 - ____ = 182

49.   78 - ____ = 40

50.   246 - ____ = 202

1.   One Hundred Ninety-Three - Fifteen = ____

2.   Two Hundred Thirty-Eight - One Hundred Twenty-Five
      = ____

3.   Two Hundred Eighty-Two - Two Hundred Eighteen =
      ____

4.   Two Hundred Fifty-One - Thirty-Nine = ____

5.   One Hundred Seventy-One - One Hundred Forty-Four
      = ____

6.   Two Hundred Forty-Four - Fifty-Eight = ____

7.   Two Hundred Fourteen - Twelve = ____

8.   Seventy-Eight - Twenty-Six = ____

9.   Two Hundred Forty-Two - One Hundred Eight = ____

10.   Two Hundred Thirteen - Ninety-Two = ____

11.   Two Hundred Thirty-Seven - One Hundred Two =
      ____

12.   Two Hundred Twenty-One - Twenty-Eight = _____

13.   Two Hundred Eighty-Five - Two Hundred Seventy-Six = _____

14.   Two Hundred Sixty-Three - Two Hundred Fifty-Seven = _____

15.   Two Hundred Two - Twenty-One = _____

16.   Two Hundred Sixty-Two - Two Hundred Fifteen = _____

17.   One Hundred Eleven - Fifty-Three = _____

18.   One Hundred Twenty-Six - Seventy-Nine = _____

19.   One Hundred Twelve - Thirty-Four = _____

20.   Two Hundred Ninety-Two - One Hundred Eighty-Nine = _____

21.   One Hundred Twenty-Seven - Ten = _____

22.   One Hundred Sixty-Eight - Forty-Six = _____

23.   Two Hundred Seventy - One Hundred Fifty-Six = _____

24.   One Hundred Sixty-Nine - Six = _____

25.   One Hundred Ninety-Nine - One Hundred Forty = _____

26.   One Hundred Sixty-Nine - Ninety-Five = _____

27.	Two Hundred Eighty-Three - Two Hundred Forty-Two = ____
28.	Two Hundred Thirty-Two - Sixty-Six = ____
29.	One Hundred Ninety-Four - One Hundred Forty-Seven = ____
30.	One Hundred Eighty-Five - One Hundred Sixty-Four = ____
1.	One Hundred Ninety-Three - Fifteen = One Hundred Seventy-Eight
2.	Two Hundred Thirty-Eight - One Hundred Twenty-Five = One Hundred Thirteen
3.	Two Hundred Eighty-Two - Two Hundred Eighteen = Sixty-Four
4.	Two Hundred Fifty-One - Thirty-Nine = Two Hundred Twelve
5.	One Hundred Seventy-One - One Hundred Forty-Four = Twenty-Seven
6.	Two Hundred Forty-Four - Fifty-Eight = One Hundred Eighty-Six
7.	Two Hundred Fourteen - Twelve = Two Hundred Two
8.	Seventy-Eight - Twenty-Six = Fifty-Two
9.	Two Hundred Forty-Two - One Hundred Eight = One

	Hundred Thirty-Four
10.	Two Hundred Thirteen - Ninety-Two = One Hundred Twenty-One
11.	Two Hundred Thirty-Seven - One Hundred Two = One Hundred Thirty-Five
12.	Two Hundred Twenty-One - Twenty-Eight = One Hundred Ninety-Three
13.	Two Hundred Eighty-Five - Two Hundred Seventy-Six = Nine
14.	Two Hundred Sixty-Three - Two Hundred Fifty-Seven = Six
15.	Two Hundred Two - Twenty-One = One Hundred Eighty-One
16.	Two Hundred Sixty-Two - Two Hundred Fifteen = Forty-Seven
17.	One Hundred Eleven - Fifty-Three = Fifty-Eight
18.	One Hundred Twenty-Six - Seventy-Nine = Forty-Seven
19.	One Hundred Twelve - Thirty-Four = Seventy-Eight
20.	Two Hundred Ninety-Two - One Hundred Eighty-Nine = One Hundred Three
21.	One Hundred Twenty-Seven - Ten = One Hundred Seventeen

22.	One Hundred Sixty-Eight - Forty-Six = One Hundred Twenty-Two
23.	Two Hundred Seventy - One Hundred Fifty-Six = One Hundred Fourteen
24.	One Hundred Sixty-Nine - Six = One Hundred Sixty-Three
25.	One Hundred Ninety-Nine - One Hundred Forty = Fifty-Nine
26.	One Hundred Sixty-Nine - Ninety-Five = Seventy-Four
27.	Two Hundred Eighty-Three - Two Hundred Forty-Two = Forty-One
28.	Two Hundred Thirty-Two - Sixty-Six = One Hundred Sixty-Six
29.	One Hundred Ninety-Four - One Hundred Forty-Seven = Forty-Seven
30.	One Hundred Eighty-Five - One Hundred Sixty-Four = Twenty-One
1.	_____ - Fifteen = One Hundred Seventy-Eight
2.	_____ - One Hundred Twenty-Five = One Hundred Thirteen
3.	_____ - Two Hundred Eighteen = Sixty-Four
4.	_____ - Thirty-Nine = Two Hundred Twelve

5.   _____ - One Hundred Forty-Four = Twenty-Seven

6.   _____ - Fifty-Eight = One Hundred Eighty-Six

7.   _____ - Twelve = Two Hundred Two

8.   _____ - Twenty-Six = Fifty-Two

9.   _____ - One Hundred Eight = One Hundred Thirty-Four

10.   _____ - Ninety-Two = One Hundred Twenty-One

11.   _____ - One Hundred Two = One Hundred Thirty-Five

12.   _____ - Twenty-Eight = One Hundred Ninety-Three

13.   _____ - Two Hundred Seventy-Six = Nine

14.   _____ - Two Hundred Fifty-Seven = Six

15.   _____ - Twenty-One = One Hundred Eighty-One

16.   _____ - Two Hundred Fifteen = Forty-Seven

17.   _____ - Fifty-Three = Fifty-Eight

18.   _____ - Seventy-Nine = Forty-Seven

19.   _____ - Thirty-Four = Seventy-Eight

20.   _____ - One Hundred Eighty-Nine = One Hundred
          Three

21.   _____ - Ten = One Hundred Seventeen

22. ____ - Forty-Six = One Hundred Twenty-Two

23. ____ - One Hundred Fifty-Six = One Hundred Fourteen

24. ____ - Six = One Hundred Sixty-Three

25. ____ - One Hundred Forty = Fifty-Nine

26. ____ - Ninety-Five = Seventy-Four

27. ____ - Two Hundred Forty-Two = Forty-One

28. ____ - Sixty-Six = One Hundred Sixty-Six

29. ____ - One Hundred Forty-Seven = Forty-Seven

30. ____ - One Hundred Sixty-Four = Twenty-One

1. One Hundred Ninety-Three - ____ = One Hundred Seventy-Eight

2. Two Hundred Thirty-Eight - ____ = One Hundred Thirteen

3. Two Hundred Eighty-Two - ____ = Sixty-Four

4. Two Hundred Fifty-One - ____ = Two Hundred Twelve

5. One Hundred Seventy-One - ____ = Twenty-Seven

6. Two Hundred Forty-Four - ____ = One Hundred Eighty-Six

7. Two Hundred Fourteen - ____ = Two Hundred Two

8.   Seventy-Eight - _____ = Fifty-Two

9.   Two Hundred Forty-Two - _____ = One Hundred Thirty-Four

10.   Two Hundred Thirteen - _____ = One Hundred Twenty-One

11.   Two Hundred Thirty-Seven - _____ = One Hundred Thirty-Five

12.   Two Hundred Twenty-One - _____ = One Hundred Ninety-Three

13.   Two Hundred Eighty-Five - _____ = Nine

14.   Two Hundred Sixty-Three - _____ = Six

15.   Two Hundred Two - _____ = One Hundred Eighty-One

16.   Two Hundred Sixty-Two - _____ = Forty-Seven

17.   One Hundred Eleven - _____ = Fifty-Eight

18.   One Hundred Twenty-Six - _____ = Forty-Seven

19.   One Hundred Twelve - _____ = Seventy-Eight

20.   Two Hundred Ninety-Two - _____ = One Hundred Three

21.   One Hundred Twenty-Seven - _____ = One Hundred Seventeen

22.   One Hundred Sixty-Eight - _____ = One Hundred

	Twenty-Two
23.	Two Hundred Seventy - _____ = One Hundred Fourteen
24.	One Hundred Sixty-Nine - _____ = One Hundred Sixty-Three
25.	One Hundred Ninety-Nine - _____ = Fifty-Nine
26.	One Hundred Sixty-Nine - _____ = Seventy-Four
27.	Two Hundred Eighty-Three - _____ = Forty-One
28.	Two Hundred Thirty-Two - _____ = One Hundred Sixty-Six
29.	One Hundred Ninety-Four - _____ = Forty-Seven
30.	One Hundred Eighty-Five - _____ = Twenty-One

GEORGE WASHINGTON BRIDGE OVER THE HUDSON RIVER

CHAPTER 4: K-4 MATH EXERCISES

MULTIPLICATIONS

1. 11 x 11 = _____

2. 13 x 14 = _____

3. 20 x 9 = _____

4. 16 x 17 = _____

5. 8 x 10 = _____

6. 7 x 13 = _____

7. 16 x 17 = _____

8. 19 x 19 = _____

9. 14 x 18 = _____

10. 12 x 12 = _____

11. 19 x 13 = _____

12. 17 x 17 = _____

13. 11 x 8 = _____

14. 8 x 19 = _____

15. 17 x 17 = _____

16.   13 x 19 = _____

17.   12 x 20 = _____

18.   8 x 14 = _____

19.   10 x 11 = _____

20.   8 x 9 = _____

21.   5 x 17 = _____

22.   16 x 11 = _____

23.   8 x 10 = _____

24.   4 x 5 = _____

25.   13 x 7 = _____

26.   7 x 20 = _____

27.   14 x 4 = _____

28.   13 x 18 = _____

29.   11 x 7 = _____

30.   20 x 4 = _____

31.   17 x 7 = _____

32.   16 x 13 = _____

33. 8 x 8 = _____

34. 15 x 20 = _____

35. 19 x 15 = _____

36. 6 x 10 = _____

37. 13 x 16 = _____

38. 19 x 17 = _____

39. 6 x 11 = _____

40. 17 x 4 = _____

41. 19 x 12 = _____

42. 4 x 17 = _____

43. 5 x 13 = _____

44. 20 x 10 = _____

45. 10 x 7 = _____

46. 18 x 19 = _____

47. 9 x 15 = _____

48. 15 x 19 = _____

49. 13 x 20 = _____

50.	**5 x 7 = _____**
1.	**11 x 11 = 121**
2.	**13 x 14 = 182**
3.	**20 x 9 = 180**
4.	**16 x 17 = 272**
5.	**8 x 10 = 80**
6.	**7 x 13 = 91**
7.	**16 x 17 = 272**
8.	**19 x 19 = 361**
9.	**14 x 18 = 252**
10.	**12 x 12 = 144**
11.	**19 x 13 = 247**
12.	**17 x 17 = 289**
13.	**11 x 8 = 88**
14.	**8 x 19 = 152**
15.	**17 x 17 = 289**
16.	**13 x 19 = 247**

17.	12 x 20 = 240
18.	8 x 14 = 112
19.	10 x 11 = 110
20.	8 x 9 = 72
21.	5 x 17 = 85
22.	16 x 11 = 176
23.	8 x 10 = 80
24.	4 x 5 = 20
25.	13 x 7 = 91
26.	7 x 20 = 140
27.	14 x 4 = 56
28.	13 x 18 = 234
29.	11 x 7 = 77
30.	20 x 4 = 80
31.	17 x 7 = 119
32.	16 x 13 = 208
33.	8 x 8 = 64

34.	15 x 20 = 300
35.	19 x 15 = 285
36.	6 x 10 = 60
37.	13 x 16 = 208
38.	19 x 17 = 323
39.	6 x 11 = 66
40.	17 x 4 = 68
41.	19 x 12 = 228
42.	4 x 17 = 68
43.	5 x 13 = 65
44.	20 x 10 = 200
45.	10 x 7 = 70
46.	18 x 19 = 342
47.	9 x 15 = 135
48.	15 x 19 = 285
49.	13 x 20 = 260
50.	5 x 7 = 35

1. ____ x 11 = 121

2. ____ x 14 = 182

3. ____ x 9 = 180

4. ____ x 17 = 272

5. ____ x 10 = 80

6. ____ x 13 = 91

7. ____ x 17 = 272

8. ____ x 19 = 361

9. ____ x 18 = 252

10. ____ x 12 = 144

11. ____ x 13 = 247

12. ____ x 17 = 289

13. ____ x 8 = 88

14. ____ x 19 = 152

15. ____ x 17 = 289

16. ____ x 19 = 247

17. ____ x 20 = 240

18. _____ x 14 = 112

19. _____ x 11 = 110

20. _____ x 9 = 72

21. _____ x 17 = 85

22. _____ x 11 = 176

23. _____ x 10 = 80

24. _____ x 5 = 20

25. _____ x 7 = 91

26. _____ x 20 = 140

27. _____ x 4 = 56

28. _____ x 18 = 234

29. _____ x 7 = 77

30. _____ x 4 = 80

31. _____ x 7 = 119

32. _____ x 13 = 208

33. _____ x 8 = 64

34. _____ x 20 = 300

35. _____ x 15 = 285

36. _____ x 10 = 60

37. _____ x 16 = 208

38. _____ x 17 = 323

39. _____ x 11 = 66

40. _____ x 4 = 68

41. _____ x 12 = 228

42. _____ x 17 = 68

43. _____ x 13 = 65

44. _____ x 10 = 200

45. _____ x 7 = 70

46. _____ x 19 = 342

47. _____ x 15 = 135

48. _____ x 19 = 285

49. _____ x 20 = 260

50. _____ x 7 = 35

1. 11 x _____ = 121

2.   13 x _____ = 182

3.   20 x _____ = 180

4.   16 x _____ = 272

5.   8 x _____ = 80

6.   7 x _____ = 91

7.   16 x _____ = 272

8.   19 x _____ = 361

9.   14 x _____ = 252

10.   12 x _____ = 144

11.   19 x _____ = 247

12.   17 x _____ = 289

13.   11 x _____ = 88

14.   8 x _____ = 152

15.   17 x _____ = 289

16.   13 x _____ = 247

17.   12 x _____ = 240

18.   8 x _____ = 112

19.	10 x _____ = 110
20.	8 x _____ = 72
21.	5 x _____ = 85
22.	16 x _____ = 176
23.	8 x _____ = 80
24.	4 x _____ = 20
25.	13 x _____ = 91
26.	7 x _____ = 140
27.	14 x _____ = 56
28.	13 x _____ = 234
29.	11 x _____ = 77
30.	20 x _____ = 80
31.	17 x _____ = 119
32.	16 x _____ = 208
33.	8 x _____ = 64
34.	15 x _____ = 300
35.	19 x _____ = 285

36.   6 x _____ = 60

37.   13 x _____ = 208

38.   19 x _____ = 323

39.   6 x _____ = 66

40.   17 x _____ = 68

41.   19 x _____ = 228

42.   4 x _____ = 68

43.   5 x _____ = 65

44.   20 x _____ = 200

45.   10 x _____ = 70

46.   18 x _____ = 342

47.   9 x _____ = 135

48.   15 x _____ = 285

49.   13 x _____ = 260

50.   5 x _____ = 35

1.   Eleven x Eleven = _____

2.   Thirteen x Fourteen = _____

3.   Twenty x Nine = ____

4.   Sixteen x Seventeen = ____

5.   Eight x Ten = ____

6.   Seven x Thirteen = ____

7.   Sixteen x Seventeen = ____

8.   Nineteen x Nineteen = ____

9.   Fourteen x Eighteen = ____

10.   Twelve x Twelve = ____

11.   Nineteen x Thirteen = ____

12.   Seventeen x Seventeen = ____

13.   Eleven x Eight = ____

14.   Eight x Nineteen = ____

15.   Seventeen x Seventeen = ____

16.   Thirteen x Nineteen = ____

17.   Twelve x Twenty = ____

18.   Eight x Fourteen = ____

19.   Ten x Eleven = ____

20.   Eight x Nine = _____

21.   Five x Seventeen = _____

22.   Sixteen x Eleven = _____

23.   Eight x Ten = _____

24.   Four x Five = _____

25.   Thirteen x Seven = _____

26.   Seven x Twenty = _____

27.   Fourteen x Four = _____

28.   Thirteen x Eighteen = _____

29.   Eleven x Seven = _____

30.   Twenty x Four = _____

1.   Eleven x Eleven = One Hundred Twenty-One

2.   Thirteen x Fourteen = One Hundred Eighty-Two

3.   Twenty x Nine = One Hundred Eighty

4.   Sixteen x Seventeen = Two Hundred Seventy-Two

5.   Eight x Ten = Eighty

6.   Seven x Thirteen = Ninety-One

7.   Sixteen x Seventeen = Two Hundred Seventy-Two

8.   Nineteen x Nineteen = Three Hundred Sixty-One

9.   Fourteen x Eighteen = Two Hundred Fifty-Two

10. Twelve x Twelve = One Hundred Forty-Four

11. Nineteen x Thirteen = Two Hundred Forty-Seven

12. Seventeen x Seventeen = Two Hundred Eighty-Nine

13. Eleven x Eight = Eighty-Eight

14. Eight x Nineteen = One Hundred Fifty-Two

15. Seventeen x Seventeen = Two Hundred Eighty-Nine

16. Thirteen x Nineteen = Two Hundred Forty-Seven

17. Twelve x Twenty = Two Hundred Forty

18. Eight x Fourteen = One Hundred Twelve

19. Ten x Eleven = One Hundred Ten

20. Eight x Nine = Seventy-Two

21. Five x Seventeen = Eighty-Five

22. Sixteen x Eleven = One Hundred Seventy-Six

23. Eight x Ten = Eighty

24.	Four x Five = Twenty
25.	Thirteen x Seven = Ninety-One
26.	Seven x Twenty = One Hundred Forty
27.	Fourteen x Four = Fifty-Six
28.	Thirteen x Eighteen = Two Hundred Thirty-Four
29.	Eleven x Seven = Seventy-Seven
30.	Twenty x Four = Eighty
1.	_____ x Eleven = One Hundred Twenty-One
2.	_____ x Fourteen = One Hundred Eighty-Two
3.	_____ x Nine = One Hundred Eighty
4.	_____ x Seventeen = Two Hundred Seventy-Two
5.	_____ x Ten = Eighty
6.	_____ x Thirteen = Ninety-One
7.	_____ x Seventeen = Two Hundred Seventy-Two
8.	_____ x Nineteen = Three Hundred Sixty-One
9.	_____ x Eighteen = Two Hundred Fifty-Two
10.	_____ x Twelve = One Hundred Forty-Four

11. _____ x Thirteen = Two Hundred Forty-Seven

12. _____ x Seventeen = Two Hundred Eighty-Nine

13. _____ x Eight = Eighty-Eight

14. _____ x Nineteen = One Hundred Fifty-Two

15. _____ x Seventeen = Two Hundred Eighty-Nine

16. _____ x Nineteen = Two Hundred Forty-Seven

17. _____ x Twenty = Two Hundred Forty

18. _____ x Fourteen = One Hundred Twelve

19. _____ x Eleven = One Hundred Ten

20. _____ x Nine = Seventy-Two

21. _____ x Seventeen = Eighty-Five

22. _____ x Eleven = One Hundred Seventy-Six

23. _____ x Ten = Eighty

24. _____ x Five = Twenty

25. _____ x Seven = Ninety-One

26. _____ x Twenty = One Hundred Forty

27. _____ x Four = Fifty-Six

28.   _____ x Eighteen = Two Hundred Thirty-Four

29.   _____ x Seven = Seventy-Seven

30.   _____ x Four = Eighty

1.   Eleven x _____ = One Hundred Twenty-One

2.   Thirteen x _____ = One Hundred Eighty-Two

3.   Twenty x _____ = One Hundred Eighty

4.   Sixteen x _____ = Two Hundred Seventy-Two

5.   Eight x _____ = Eighty

6.   Seven x _____ = Ninety-One

7.   Sixteen x _____ = Two Hundred Seventy-Two

8.   Nineteen x _____ = Three Hundred Sixty-One

9.   Fourteen x _____ = Two Hundred Fifty-Two

10.   Twelve x _____ = One Hundred Forty-Four

11.   Nineteen x _____ = Two Hundred Forty-Seven

12.   Seventeen x _____ = Two Hundred Eighty-Nine

13.   Eleven x _____ = Eighty-Eight

14.   Eight x _____ = One Hundred Fifty-Two

15. Seventeen x _____ = Two Hundred Eighty-Nine

16. Thirteen x _____ = Two Hundred Forty-Seven

17. Twelve x _____ = Two Hundred Forty

18. Eight x _____ = One Hundred Twelve

19. Ten x _____ = One Hundred Ten

20. Eight x _____ = Seventy-Two

21. Five x _____ = Eighty-Five

22. Sixteen x _____ = One Hundred Seventy-Six

23. Eight x _____ = Eighty

24. Four x _____ = Twenty

25. Thirteen x _____ = Ninety-One

26. Seven x _____ = One Hundred Forty

27. Fourteen x _____ = Fifty-Six

28. Thirteen x _____ = Two Hundred Thirty-Four

29. Eleven x _____ = Seventy-Seven

30. Twenty x _____ = Eighty

TAIL VIEW FROM B-17 SUPERFORTRESS OVER NEW YORK CITY

CHAPTER 4: K-4 MATH EXERCISES

DIVISIONS

1.	$75 \div 5 =$ _____
2.	$91 \div 7 =$ _____
3.	$70 \div 35 =$ _____
4.	$84 \div 14 =$ _____
5.	$90 \div 10 =$ _____
6.	$40 \div 8 =$ _____
7.	$28 \div 14 =$ _____
8.	$90 \div 6 =$ _____
9.	$44 \div 11 =$ _____
10.	$72 \div 4 =$ _____
11.	$12 \div 6 =$ _____
12.	$70 \div 10 =$ _____
13.	$76 \div 4 =$ _____
14.	$51 \div 17 =$ _____
15.	$65 \div 13 =$ _____

16. $20 \div 10 =$ ____

17. $81 \div 9 =$ ____

18. $28 \div 7 =$ ____

19. $44 \div 22 =$ ____

20. $84 \div 6 =$ ____

21. $72 \div 18 =$ ____

22. $92 \div 46 =$ ____

23. $64 \div 8 =$ ____

24. $50 \div 10 =$ ____

25. $84 \div 14 =$ ____

26. $52 \div 13 =$ ____

27. $98 \div 14 =$ ____

28. $52 \div 13 =$ ____

29. $49 \div 7 =$ ____

30. $30 \div 5 =$ ____

31. $26 \div 13 =$ ____

32. $44 \div 4 =$ ____

33. $88 \div 22 =$ _____

34. $93 \div 31 =$ _____

35. $86 \div 43 =$ _____

36. $20 \div 5 =$ _____

37. $84 \div 42 =$ _____

38. $12 \div 6 =$ _____

39. $48 \div 4 =$ _____

40. $45 \div 5 =$ _____

41. $56 \div 14 =$ _____

42. $72 \div 8 =$ _____

43. $99 \div 9 =$ _____

44. $45 \div 15 =$ _____

45. $49 \div 7 =$ _____

46. $45 \div 9 =$ _____

47. $65 \div 13 =$ _____

48. $80 \div 8 =$ _____

49. $74 \div 37 =$ _____

50.	$40 \div 20 = \underline{\qquad}$
1.	$75 \div 5 = 15$
2.	$91 \div 7 = 13$
3.	$70 \div 35 = 2$
4.	$84 \div 14 = 6$
5.	$90 \div 10 = 9$
6.	$40 \div 8 = 5$
7.	$28 \div 14 = 2$
8.	$90 \div 6 = 15$
9.	$44 \div 11 = 4$
10.	$72 \div 4 = 18$
11.	$12 \div 6 = 2$
12.	$70 \div 10 = 7$
13.	$76 \div 4 = 19$
14.	$51 \div 17 = 3$
15.	$65 \div 13 = 5$
16.	$20 \div 10 = 2$

17.	$81 \div 9 = 9$
18.	$28 \div 7 = 4$
19.	$44 \div 22 = 2$
20.	$84 \div 6 = 14$
21.	$72 \div 18 = 4$
22.	$92 \div 46 = 2$
23.	$64 \div 8 = 8$
24.	$50 \div 10 = 5$
25.	$84 \div 14 = 6$
26.	$52 \div 13 = 4$
27.	$98 \div 14 = 7$
28.	$52 \div 13 = 4$
29.	$49 \div 7 = 7$
30.	$30 \div 5 = 6$
31.	$26 \div 13 = 2$
32.	$44 \div 4 = 11$
33.	$88 \div 22 = 4$

34.	$93 \div 31 = 3$
35.	$86 \div 43 = 2$
36.	$20 \div 5 = 4$
37.	$84 \div 42 = 2$
38.	$12 \div 6 = 2$
39.	$48 \div 4 = 12$
40.	$45 \div 5 = 9$
41.	$56 \div 14 = 4$
42.	$72 \div 8 = 9$
43.	$99 \div 9 = 11$
44.	$45 \div 15 = 3$
45.	$49 \div 7 = 7$
46.	$45 \div 9 = 5$
47.	$65 \div 13 = 5$
48.	$80 \div 8 = 10$
49.	$74 \div 37 = 2$
50.	$40 \div 20 = 2$

1. _____ ÷ 5 = 15

2. _____ ÷ 7 = 13

3. _____ ÷ 35 = 2

4. _____ ÷ 14 = 6

5. _____ ÷ 10 = 9

6. _____ ÷ 8 = 5

7. _____ ÷ 14 = 2

8. _____ ÷ 6 = 15

9. _____ ÷ 11 = 4

10. _____ ÷ 4 = 18

11. _____ ÷ 6 = 2

12. _____ ÷ 10 = 7

13. _____ ÷ 4 = 19

14. _____ ÷ 17 = 3

15. _____ ÷ 13 = 5

16. _____ ÷ 10 = 2

17. _____ ÷ 9 = 9

18. _____ ÷ 7 = 4

19. _____ ÷ 22 = 2

20. _____ ÷ 6 = 14

21. _____ ÷ 18 = 4

22. _____ ÷ 46 = 2

23. _____ ÷ 8 = 8

24. _____ ÷ 10 = 5

25. _____ ÷ 14 = 6

26. _____ ÷ 13 = 4

27. _____ ÷ 14 = 7

28. _____ ÷ 13 = 4

29. _____ ÷ 7 = 7

30. _____ ÷ 5 = 6

31. _____ ÷ 13 = 2

32. _____ ÷ 4 = 11

33. _____ ÷ 22 = 4

34. _____ ÷ 31 = 3

35.	___ ÷ 43 = 2
36.	___ ÷ 5 = 4
37.	___ ÷ 42 = 2
38.	___ ÷ 6 = 2
39.	___ ÷ 4 = 12
40.	___ ÷ 5 = 9
41.	___ ÷ 14 = 4
42.	___ ÷ 8 = 9
43.	___ ÷ 9 = 11
44.	___ ÷ 15 = 3
45.	___ ÷ 7 = 7
46.	___ ÷ 9 = 5
47.	___ ÷ 13 = 5
48.	___ ÷ 8 = 10
49.	___ ÷ 37 = 2
50.	___ ÷ 20 = 2
1.	75 ÷ ___ = 15

2. $91 \div \underline{\hphantom{00}} = 13$

3. $70 \div \underline{\hphantom{00}} = 2$

4. $84 \div \underline{\hphantom{00}} = 6$

5. $90 \div \underline{\hphantom{00}} = 9$

6. $40 \div \underline{\hphantom{00}} = 5$

7. $28 \div \underline{\hphantom{00}} = 2$

8. $90 \div \underline{\hphantom{00}} = 15$

9. $44 \div \underline{\hphantom{00}} = 4$

10. $72 \div \underline{\hphantom{00}} = 18$

11. $12 \div \underline{\hphantom{00}} = 2$

12. $70 \div \underline{\hphantom{00}} = 7$

13. $76 \div \underline{\hphantom{00}} = 19$

14. $51 \div \underline{\hphantom{00}} = 3$

15. $65 \div \underline{\hphantom{00}} = 5$

16. $20 \div \underline{\hphantom{00}} = 2$

17. $81 \div \underline{\hphantom{00}} = 9$

18. $28 \div \underline{\hphantom{00}} = 4$

19.　　44 ÷ _____ = 2

20.　　84 ÷ _____ = 14

21.　　72 ÷ _____ = 4

22.　　92 ÷ _____ = 2

23.　　64 ÷ _____ = 8

24.　　50 ÷ _____ = 5

25.　　84 ÷ _____ = 6

26.　　52 ÷ _____ = 4

27.　　98 ÷ _____ = 7

28.　　52 ÷ _____ = 4

29.　　49 ÷ _____ = 7

30.　　30 ÷ _____ = 6

31.　　26 ÷ _____ = 2

32.　　44 ÷ _____ = 11

33.　　88 ÷ _____ = 4

34.　　93 ÷ _____ = 3

35.　　86 ÷ _____ = 2

36. $20 \div$ _____ $= 4$

37. $84 \div$ _____ $= 2$

38. $12 \div$ _____ $= 2$

39. $48 \div$ _____ $= 12$

40. $45 \div$ _____ $= 9$

41. $56 \div$ _____ $= 4$

42. $72 \div$ _____ $= 9$

43. $99 \div$ _____ $= 11$

44. $45 \div$ _____ $= 3$

45. $49 \div$ _____ $= 7$

46. $45 \div$ _____ $= 5$

47. $65 \div$ _____ $= 5$

48. $80 \div$ _____ $= 10$

49. $74 \div$ _____ $= 2$

50. $40 \div$ _____ $= 2$

1.   **Seventy-Five ÷ Five = _____**

2.   **Ninety-One ÷ Seven = _____**

3.   Seventy ÷ Thirty-Five = _____

4.   Eighty-Four ÷ Fourteen = _____

5.   Ninety ÷ Ten = _____

6.   Forty ÷ Eight = _____

7.   Twenty-Eight ÷ Fourteen = _____

8.   Ninety ÷ Six = _____

9.   Forty-Four ÷ Eleven = _____

10.   Seventy-Two ÷ Four = _____

11.   Twelve ÷ Six = _____

12.   Seventy ÷ Ten = _____

13.   Seventy-Six ÷ Four = _____

14.   Fifty-One ÷ Seventeen = _____

15.   Sixty-Five ÷ Thirteen = _____

16.   Twenty ÷ Ten = _____

17.   Eighty-One ÷ Nine = _____

18.   Twenty-Eight ÷ Seven = _____

19.   Forty-Four ÷ Twenty-Two = _____

20.	Eighty-Four ÷ Six = _____
21.	Seventy-Two ÷ Eighteen = _____
22.	Ninety-Two ÷ Forty-Six = _____
23.	Sixty-Four ÷ Eight = _____
24.	Fifty ÷ Ten = _____
25.	Eighty-Four ÷ Fourteen = _____
26.	Fifty-Two ÷ Thirteen = _____
27.	Ninety-Eight ÷ Fourteen = _____
28.	Fifty-Two ÷ Thirteen = _____
29.	Forty-Nine ÷ Seven = _____
30.	Thirty ÷ Five = _____
1.	Seventy-Five ÷ Five = Fifteen
2.	Ninety-One ÷ Seven = Thirteen
3.	Seventy ÷ Thirty-Five = Two
4.	Eighty-Four ÷ Fourteen = Six
5.	Ninety ÷ Ten = Nine
6.	Forty ÷ Eight = Five

7.	Twenty-Eight ÷ Fourteen = Two
8.	Ninety ÷ Six = Fifteen
9.	Forty-Four ÷ Eleven = Four
10.	Seventy-Two ÷ Four = Eighteen
11.	Twelve ÷ Six = Two
12.	Seventy ÷ Ten = Seven
13.	Seventy-Six ÷ Four = Nineteen
14.	Fifty-One ÷ Seventeen = Three
15.	Sixty-Five ÷ Thirteen = Five
16.	Twenty ÷ Ten = Two
17.	Eighty-One ÷ Nine = Nine
18.	Twenty-Eight ÷ Seven = Four
19.	Forty-Four ÷ Twenty-Two = Two
20.	Eighty-Four ÷ Six = Fourteen
21.	Seventy-Two ÷ Eighteen = Four
22.	Ninety-Two ÷ Forty-Six = Two
23.	Sixty-Four ÷ Eight = Eight

24.	Fifty ÷ Ten = Five
25.	Eighty-Four ÷ Fourteen = Six
26.	Fifty-Two ÷ Thirteen = Four
27.	Ninety-Eight ÷ Fourteen = Seven
28.	Fifty-Two ÷ Thirteen = Four
29.	Forty-Nine ÷ Seven = Seven
30.	Thirty ÷ Five = Six
1.	_____ ÷ Five = Fifteen
2.	_____ ÷ Seven = Thirteen
3.	_____ ÷ Thirty-Five = Two
4.	_____ ÷ Fourteen = Six
5.	_____ ÷ Ten = Nine
6.	_____ ÷ Eight = Five
7.	_____ ÷ Fourteen = Two
8.	_____ ÷ Six = Fifteen
9.	_____ ÷ Eleven = Four
10.	_____ ÷ Four = Eighteen

11.	____ ÷ Six = Two
12.	____ ÷ Ten = Seven
13.	____ ÷ Four = Nineteen
14.	____ ÷ Seventeen = Three
15.	____ ÷ Thirteen = Five
16.	____ ÷ Ten = Two
17.	____ ÷ Nine = Nine
18.	____ ÷ Seven = Four
19.	____ ÷ Twenty-Two = Two
20.	____ ÷ Six = Fourteen
21.	____ ÷ Eighteen = Four
22.	____ ÷ Forty-Six = Two
23.	____ ÷ Eight = Eight
24.	____ ÷ Ten = Five
25.	____ ÷ Fourteen = Six
26.	____ ÷ Thirteen = Four
27.	____ ÷ Fourteen = Seven

28.	____ ÷ Thirteen = Four
29.	____ ÷ Seven = Seven
30.	____ ÷ Five = Six
1.	Seventy-Five ÷ ____ = Fifteen
2.	Ninety-One ÷ ____ = Thirteen
3.	Seventy ÷ ____ = Two
4.	Eighty-Four ÷ ____ = Six
5.	Ninety ÷ ____ = Nine
6.	Forty ÷ ____ = Five
7.	Twenty-Eight ÷ ____ = Two
8.	Ninety ÷ ____ = Fifteen
9.	Forty-Four ÷ ____ = Four
10.	Seventy-Two ÷ ____ = Eighteen
11.	Twelve ÷ ____ = Two
12.	Seventy ÷ ____ = Seven
13.	Seventy-Six ÷ ____ = Nineteen
14.	Fifty-One ÷ ____ = Three

15. Sixty-Five ÷ _____ = Five

16. Twenty ÷ _____ = Two

17. Eighty-One ÷ _____ = Nine

18. Twenty-Eight ÷ _____ = Four

19. Forty-Four ÷ _____ = Two

20. Eighty-Four ÷ _____ = Fourteen

21. Seventy-Two ÷ _____ = Four

22. Ninety-Two ÷ _____ = Two

23. Sixty-Four ÷ _____ = Eight

24. Fifty ÷ _____ = Five

25. Eighty-Four ÷ _____ = Six

26. Fifty-Two ÷ _____ = Four

27. Ninety-Eight ÷ _____ = Seven

28. Fifty-Two ÷ _____ = Four

29. Forty-Nine ÷ _____ = Seven

30. Thirty ÷ _____ = Six

WORLD TRADE CENTER 7 CORNER VIEW IN NYC

CHAPTER 4: K-4 MATH EXERCISES

MIXED OPERATIONS - SERIES 1

Computer Science operation symbols are used for multiplication (*) and division(/).

1.	(27 - 24) * 7 = _____
2.	25 + (20 - 18) = _____
3.	23 + 13 + 11 = _____
4.	25 + (23 - 13) = _____
5.	28 + 17 + 14 = _____
6.	26 * 8 + 3 = _____
7.	8 * 7 + 6 = _____
8.	(22 - 15) * 10 = _____
9.	(27 - 23) + 6 = _____
10.	24 - 20 + 4 = _____
11.	(25 + 22) + 16 = _____
12.	26 - 16 + 9 = _____
13.	29 + 26 + 13 = _____
14.	(28 * 10) - 8 = _____

15. $(24 - 18) * 14 = $ _____

16. $26 + 25 - 12 = $ _____

17. $(10 + 9) + 7 = $ _____

18. $(17 + 11) + 8 = $ _____

19. $29 * (28 - 19) = $ _____

20. $(16 - 15) * 9 = $ _____

21. $(26 - 10) - 8 = $ _____

22. $(29 + 28) + 16 = $ _____

23. $27 - 23 + 18 = $ _____

24. $(16 + 13) * 6 = $ _____

25. $28 * 8 / 7 = $ _____

26. $(30 - 25) * 24 = $ _____

27. $(24 - 22) * 11 = $ _____

28. $(22 - 19) * 14 = $ _____

29. $(28 + 23) + 10 = $ _____

30. $27 + (22 - 13) = $ _____

31. $26 - 13 + 12 = $ _____

32.	21 - (19 - 16) = _____
33.	22 + 21 + 20 = _____
34.	14 * 6 / 4 = _____
35.	29 + (26 + 24) = _____
36.	(7 - 6) + 4 = _____
37.	(25 + 22) - 9 = _____
38.	19 - 18 + 14 = _____
39.	(18 * 14) / 7 = _____
40.	28 - 20 + 15 = _____
41.	22 - 16 + 13 = _____
42.	18 * 17 - 9 = _____
43.	17 + (10 - 5) = _____
44.	(19 - 14) + 11 = _____
45.	22 * 21 / 3 = _____
46.	28 + 27 + 24 = _____
47.	(16 * 7) + 4 = _____
48.	25 + 18 + 5 = _____

49.   28 / (25 - 23) = _____

50.   (20 * 12) - 11 = _____

51.   (27 + 26) + 9 = _____

52.   (30 + 21) + 19 = _____

53.   (28 + 17) - 13 = _____

54.   20 * 18 / 15 = _____

55.   (25 + 20) - 16 = _____

56.   (21 + 11) + 6 = _____

57.   (25 + 17) + 13 = _____

58.   (26 + 22) - 15 = _____

59.   24 - (19 - 17) = _____

60.   (28 * 12) / 3 = _____

61.   16 + (8 + 6) = _____

62.   22 + 19 - 15 = _____

63.   (29 - 27) + 10 = _____

64.   30 + (19 * 9) = _____

65.   (26 + 21) * 4 = _____

66.	(30 + 25) - 24 = _____
67.	(28 - 16) + 12 = _____
68.	14 + 11 - 6 = _____
69.	(22 + 20) + 8 = _____
70.	(23 - 17) + 12 = _____
71.	(18 * 13) + 5 = _____
72.	16 - 11 + 10 = _____
73.	27 + (19 + 15) = _____
74.	(23 - 21) * 10 = _____
75.	(25 - 18) * 6 = _____
76.	28 + 23 + 8 = _____
77.	22 + 18 * 12 = _____
78.	30 + (29 + 24) = _____
79.	(24 * 12) / 9 = _____
80.	(25 + 17) * 7 = _____
81.	(21 - 20) * 18 = _____
82.	26 + 17 + 9 = _____

83.	30 * (23 - 22) = _____
84.	(23 - 16) + 15 = _____
85.	22 * (17 - 11) = _____
86.	(18 + 12) - 8 = _____
87.	22 + (21 - 16) = _____
88.	23 + (17 - 4) = _____
89.	(21 * 13) - 9 = _____
90.	23 + (18 + 6) = _____
91.	28 + 14 * 10 = _____
92.	(30 + 29) - 24 = _____
93.	(26 + 19) + 11 = _____
94.	23 - (11 - 10) = _____
95.	29 - 10 + 8 = _____
96.	(16 + 6) + 3 = _____
97.	(30 + 23) + 13 = _____
98.	(22 + 15) + 14 = _____
99.	21 + 14 + 9 = _____

100.	(28 - 27) + 21 = _____
1.	(27 - 24) * 7 = 21
2.	25 + (20 - 18) = 27
3.	23 + 13 + 11 = 47
4.	25 + (23 - 13) = 35
5.	28 + 17 + 14 = 59
6.	26 * 8 + 3 = 211
7.	8 * 7 + 6 = 62
8.	(22 - 15) * 10 = 70
9.	(27 - 23) + 6 = 10
10.	24 - 20 + 4 = 8
11.	(25 + 22) + 16 = 63
12.	26 - 16 + 9 = 19
13.	29 + 26 + 13 = 68
14.	(28 * 10) - 8 = 272
15.	(24 - 18) * 14 = 84
16.	26 + 25 - 12 = 39

17.	$(10 + 9) + 7 = 26$
18.	$(17 + 11) + 8 = 36$
19.	$29 * (28 - 19) = 261$
20.	$(16 - 15) * 9 = 9$
21.	$(26 - 10) - 8 = 8$
22.	$(29 + 28) + 16 = 73$
23.	$27 - 23 + 18 = 22$
24.	$(16 + 13) * 6 = 174$
25.	$28 * 8 / 7 = 32$
26.	$(30 - 25) * 24 = 120$
27.	$(24 - 22) * 11 = 22$
28.	$(22 - 19) * 14 = 42$
29.	$(28 + 23) + 10 = 61$
30.	$27 + (22 - 13) = 36$
31.	$26 - 13 + 12 = 25$
32.	$21 - (19 - 16) = 18$
33.	$22 + 21 + 20 = 63$

34.	$14 * 6 / 4 = 21$
35.	$29 + (26 + 24) = 79$
36.	$(7 - 6) + 4 = 5$
37.	$(25 + 22) - 9 = 38$
38.	$19 - 18 + 14 = 15$
39.	$(18 * 14) / 7 = 36$
40.	$28 - 20 + 15 = 23$
41.	$22 - 16 + 13 = 19$
42.	$18 * 17 - 9 = 297$
43.	$17 + (10 - 5) = 22$
44.	$(19 - 14) + 11 = 16$
45.	$22 * 21 / 3 = 154$
46.	$28 + 27 + 24 = 79$
47.	$(16 * 7) + 4 = 116$
48.	$25 + 18 + 5 = 48$
49.	$28 / (25 - 23) = 14$
50.	$(20 * 12) - 11 = 229$

51.	(27 + 26) + 9 = 62
52.	(30 + 21) + 19 = 70
53.	(28 + 17) - 13 = 32
54.	20 * 18 / 15 = 24
55.	(25 + 20) - 16 = 29
56.	(21 + 11) + 6 = 38
57.	(25 + 17) + 13 = 55
58.	(26 + 22) - 15 = 33
59.	24 - (19 - 17) = 22
60.	(28 * 12) / 3 = 112
61.	16 + (8 + 6) = 30
62.	22 + 19 - 15 = 26
63.	(29 - 27) + 10 = 12
64.	30 + (19 * 9) = 201
65.	(26 + 21) * 4 = 188
66.	(30 + 25) - 24 = 31
67.	(28 - 16) + 12 = 24

68.	14 + 11 - 6 = 19
69.	(22 + 20) + 8 = 50
70.	(23 - 17) + 12 = 18
71.	(18 * 13) + 5 = 239
72.	16 - 11 + 10 = 15
73.	27 + (19 + 15) = 61
74.	(23 - 21) * 10 = 20
75.	(25 - 18) * 6 = 42
76.	28 + 23 + 8 = 59
77.	22 + 18 * 12 = 238
78.	30 + (29 + 24) = 83
79.	(24 * 12) / 9 = 32
80.	(25 + 17) * 7 = 294
81.	(21 - 20) * 18 = 18
82.	26 + 17 + 9 = 52
83.	30 * (23 - 22) = 30
84.	(23 - 16) + 15 = 22

85.	22 * (17 - 11) = 132
86.	(18 + 12) - 8 = 22
87.	22 + (21 - 16) = 27
88.	23 + (17 - 4) = 36
89.	(21 * 13) - 9 = 264
90.	23 + (18 + 6) = 47
91.	28 + 14 * 10 = 168
92.	(30 + 29) - 24 = 35
93.	(26 + 19) + 11 = 56
94.	23 - (11 - 10) = 22
95.	29 - 10 + 8 = 27
96.	(16 + 6) + 3 = 25
97.	(30 + 23) + 13 = 66
98.	(22 + 15) + 14 = 51
99.	21 + 14 + 9 = 44
100.	(28 - 27) + 21 = 22
1.	(_____ - 24) * 7 = 21

2.	_____ + (20 - 18) = 27
3.	_____ + 13 + 11 = 47
4.	25 + (23 - _____) = 35
5.	_____ + 17 + 14 = 59
6.	26 * 8 + _____ = 211
7.	8 * _____ + 6 = 62
8.	(22 - 15) * _____ = 70
9.	(27 - 23) + _____ = 10
10.	24 - _____ + 4 = 8
11.	(25 + _____) + 16 = 63
12.	26 - 16 + _____ = 19
13.	29 + 26 + _____ = 68
14.	(_____ * 10) - 8 = 272
15.	(24 - _____) * 14 = 84
16.	26 + _____ - 12 = 39
17.	(10 + 9) + _____ = 26
18.	(17 + _____) + 8 = 36

19.	_____ * (28 - 19) = 261
20.	(16 - _____) * 9 = 9
21.	(_____ - 10) - 8 = 8
22.	(29 + _____) + 16 = 73
23.	27 - 23 + _____ = 22
24.	(16 + _____) * 6 = 174
25.	_____ * 8 / 7 = 32
26.	(30 - 25) * _____ = 120
27.	(_____ - 22) * 11 = 22
28.	(22 - _____) * 14 = 42
29.	(28 + 23) + _____ = 61
30.	_____ + (22 - 13) = 36
31.	26 - 13 + _____ = 25
32.	21 - (_____ - 16) = 18
33.	22 + 21 + _____ = 63
34.	14 * _____ / 4 = 21
35.	29 + (26 + _____) = 79

36.	(7 - 6) + _____ = 5
37.	(_____ + 22) - 9 = 38
38.	19 - _____ + 14 = 15
39.	(_____ * 14) / 7 = 36
40.	28 - 20 + _____ = 23
41.	22 - _____ + 13 = 19
42.	_____ * 17 - 9 = 297
43.	17 + (10 - _____) = 22
44.	(_____ - 14) + 11 = 16
45.	_____ * 21 / 3 = 154
46.	28 + _____ + 24 = 79
47.	(_____ * 7) + 4 = 116
48.	25 + _____ + 5 = 48
49.	_____ / (25 - 23) = 14
50.	(20 * 12) - _____ = 229
51.	(27 + 26) + _____ = 62
52.	(_____ + 21) + 19 = 70

53.	$(28 + 17) -$ _____ $= 32$
54.	$20 *$ _____ $/ 15 = 24$
55.	$(25 + 20) -$ _____ $= 29$
56.	$(21 + 11) +$ _____ $= 38$
57.	$(25 + 17) +$ _____ $= 55$
58.	$($_____ $+ 22) - 15 = 33$
59.	$24 - (19 -$ _____ $) = 22$
60.	$($_____ $* 12) / 3 = 112$
61.	_____ $+ (8 + 6) = 30$
62.	_____ $+ 19 - 15 = 26$
63.	$(29 -$ _____ $) + 10 = 12$
64.	_____ $+ (19 * 9) = 201$
65.	$(26 +$ _____ $) * 4 = 188$
66.	$($_____ $+ 25) - 24 = 31$
67.	$(28 - 16) +$ _____ $= 24$
68.	_____ $+ 11 - 6 = 19$
69.	$($_____ $+ 20) + 8 = 50$

70.	(_____ - 17) + 12 = 18
71.	(18 * 13) + _____ = 239
72.	16 - _____ + 10 = 15
73.	_____ + (19 + 15) = 61
74.	(23 - 21) * _____ = 20
75.	(_____ - 18) * 6 = 42
76.	28 + 23 + _____ = 59
77.	22 + 18 * _____ = 238
78.	_____ + (29 + 24) = 83
79.	(24 * 12) / _____ = 32
80.	(25 + 17) * _____ = 294
81.	(21 - _____) * 18 = 18
82.	26 + _____ + 9 = 52
83.	30 * (_____ - 22) = 30
84.	(_____ - 16) + 15 = 22
85.	22 * (_____ - 11) = 132
86.	(18 + 12) - _____ = 22

87.	_____ + (21 - 16) = 27
88.	23 + (_____ - 4) = 36
89.	(21 * _____) - 9 = 264
90.	23 + (_____ + 6) = 47
91.	_____ + 14 * 10 = 168
92.	(_____ + 29) - 24 = 35
93.	(26 + _____) + 11 = 56
94.	23 - (_____ - 10) = 22
95.	29 - _____ + 8 = 27
96.	(16 + _____) + 3 = 25
97.	(_____ + 23) + 13 = 66
98.	(22 + 15) + _____ = 51
99.	21 + 14 + _____ = 44
100.	(28 - _____) + 21 = 22

PLAZA WITH RED BALLOON FLOWER SCULPTURE BY FRONT OF WTC 7 IN NYC

CHAPTER 4: K-4 MATH EXERCISES

MIXED OPERATIONS - SERIES 2

The parentheses can be removed in the following expressions and the calculation can be carried out in such an order which avoids fraction calculations.

1.	(22 / 20) * 10 = _____
2.	18 * (13 / 9) = _____
3.	(21 / 18) * 12 = _____
4.	(30 / 24) * 20 = _____
5.	20 * (18 / 15) = _____
6.	(27 / 15) * 10 = _____
7.	24 * (14 / 6) = _____
8.	(26 / 18) * 9 = _____
9.	16 * (9 / 6) = _____
10.	26 / 4 * 2 = _____
11.	27 * (20 / 18) = _____
12.	(22 / 16) * 8 = _____
13.	(25 / 20) * 4 = _____
14.	(30 / 18) * 9 = _____

15. $25 * (16 / 10) =$ _____

16. $18 / 4 * 2 =$ _____

17. $(26 / 10) * 5 =$ _____

18. $22 * (18 / 4) =$ _____

19. $30 / 27 * 18 =$ _____

20. $27 / 15 * 10 =$ _____

21. $28 / 16 * 12 =$ _____

22. $10 * (7 / 2) =$ _____

23. $24 / 16 * 14 =$ _____

24. $30 * (9 / 6) =$ _____

25. $25 / 15 * 12 =$ _____

26. $20 / (10 / 8) =$ _____

27. $30 / (24 / 20) =$ _____

28. $(12 / 10) * 5 =$ _____

29. $26 / 14 * 7 =$ _____

30. $(28 / 21) * 9 =$ _____

31. $12 * (10 / 8) =$ _____

32.	27 / (21 / 14) = _____
33.	28 / 8 * 4 = _____
34.	27 * (7 / 3) = _____
35.	24 * (22 / 16) = _____
36.	28 * (17 / 14) = _____
37.	30 / (20 / 18) = _____
38.	28 * (18 / 12) = _____
39.	27 / 15 * 10 = _____
40.	22 * (17 / 11) = _____
41.	(18 / 12) * 8 = _____
42.	30 / 25 * 5 = _____
43.	(28 / 21) * 6 = _____
44.	18 * (5 / 3) = _____
45.	(25 / 10) * 8 = _____
46.	24 / 9 * 6 = _____
47.	27 * (23 / 9) = _____
48.	28 * (17 / 14) = _____

49.	12 * (10 / 6) = ____
50.	24 * (21 / 14) = ____
1.	(22 / 20) * 10 = 11
2.	18 * (13 / 9) = 26
3.	(21 / 18) * 12 = 14
4.	(30 / 24) * 20 = 25
5.	20 * (18 / 15) = 24
6.	(27 / 15) * 10 = 18
7.	24 * (14 / 6) = 56
8.	(26 / 18) * 9 = 13
9.	16 * (9 / 6) = 24
10.	26 / 4 * 2 = 13
11.	27 * (20 / 18) = 30
12.	(22 / 16) * 8 = 11
13.	(25 / 20) * 4 = 5
14.	(30 / 18) * 9 = 15
15.	25 * (16 / 10) = 40

16.   18 / 4 * 2 = 9

17.   (26 / 10) * 5 = 13

18.   22 * (18 / 4) = 99

19.   30 / 27 * 18 = 20

20.   27 / 15 * 10 = 18

21.   28 / 16 * 12 = 21

22.   10 * (7 / 2) = 35

23.   24 / 16 * 14 = 21

24.   30 * (9 / 6) = 45

25.   25 / 15 * 12 = 20

26.   20 / (10 / 8) = 16

27.   30 / (24 / 20) = 25

28.   (12 / 10) * 5 = 6

29.   26 / 14 * 7 = 13

30.   (28 / 21) * 9 = 12

31.   12 * (10 / 8) = 15

32.   27 / (21 / 14) = 18

33.	28 / 8 * 4 = 14
34.	27 * (7 / 3) = 63
35.	24 * (22 / 16) = 33
36.	28 * (17 / 14) = 34
37.	30 / (20 / 18) = 27
38.	28 * (18 / 12) = 42
39.	27 / 15 * 10 = 18
40.	22 * (17 / 11) = 34
41.	(18 / 12) * 8 = 12
42.	30 / 25 * 5 = 6
43.	(28 / 21) * 6 = 8
44.	18 * (5 / 3) = 30
45.	(25 / 10) * 8 = 20
46.	24 / 9 * 6 = 16
47.	27 * (23 / 9) = 69
48.	28 * (17 / 14) = 34
49.	12 * (10 / 6) = 20

50.	24 * (21 / 14) = 36
1.	(22 / _____) * 10 = 11
2.	18 * (_____ / 9) = 26
3.	(21 / 18) * _____ = 14
4.	(30 / 24) * _____ = 25
5.	_____ * (18 / 15) = 24
6.	(27 / 15) * _____ = 18
7.	_____ * (14 / 6) = 56
8.	(26 / _____) * 9 = 13
9.	16 * (9 / _____) = 24
10.	26 / _____ * 2 = 13
11.	27 * (_____ / 18) = 30
12.	(22 / 16) * _____ = 11
13.	(25 / _____) * 4 = 5
14.	(_____ / 18) * 9 = 15
15.	25 * (16 / _____) = 40
16.	18 / 4 * _____ = 9

17.	(_____ / 10) * 5 = 13
18.	_____ * (18 / 4) = 99
19.	_____ / 27 * 18 = 20
20.	_____ / 15 * 10 = 18
21.	_____ / 16 * 12 = 21
22.	10 * (7 / _____) = 35
23.	24 / 16 * _____ = 21
24.	_____ * (9 / 6) = 45
25.	25 / _____ * 12 = 20
26.	_____ / (10 / 8) = 16
27.	_____ / (24 / 20) = 25
28.	(12 / 10) * _____ = 6
29.	_____ / 14 * 7 = 13
30.	(28 / _____) * 9 = 12
31.	12 * (_____ / 8) = 15
32.	27 / (_____ / 14) = 18
33.	_____ / 8 * 4 = 14

34.	27 * (_____ / 3) = 63
35.	24 * (22 / _____) = 33
36.	_____ * (17 / 14) = 34
37.	30 / (20 / _____) = 27
38.	28 * (18 / _____) = 42
39.	27 / 15 * _____ = 18
40.	_____ * (17 / 11) = 34
41.	(18 / _____) * 8 = 12
42.	30 / 25 * _____ = 6
43.	(28 / _____) * 6 = 8
44.	18 * (5 / _____) = 30
45.	(25 / _____) * 8 = 20
46.	24 / 9 * _____ = 16
47.	27 * (23 / _____) = 69
48.	_____ * (17 / 14) = 34
49.	12 * (_____ / 6) = 20
50.	24 * (_____ / 14) = 36

ROCKEFELLER CENTER CHRISTMAS TREE IN NYC

CHAPTER 4: K-4 MATH EXERCISES

SHOPPING MATH PROBLEMS

1.	Shopper buys 2 Pairs of Shorts at $20 and 2 T-Shirts at $8. How much is due if the sales tax is 8% ?
2.	Shopper buys 1 Pair of Espadrilles at $___ and 3 Cameras at $12. Shopper pays $100 and received the change of $47. How much is the missing unit price?
3.	Shopper buys 3 Containers at $20 and 2 Jackets at $___. Difference between the most expensive item and least expensive item is $8. How much is the missing unit price? [1] $10 [2] $14 [3] $9 [4] $12
4.	Shopper buys 1 Watch at $___, 2 Nut Drivers at $8 and 3 Radios at $16. Total purchase is $75. How much is the missing unit price?
5.	Shopper buys 1 T-Shirt at $8, 1 Pair of Sandals at $15 and 2 Pairs of Gloves at $___. Shopper pays $50 and received the change of $11. How much is the missing unit price?

1. srallod -ytxiS 2. srallod neetneveS 3. ruoF 4. srallod nevelE 5. srallod thgiE

6.	Shopper buys 1 Radio at $19, 2 T-Shirts at $8 and 2 Table Linen Sets at $14. How many total items in the shopping cart?
7.	Shopper buys 3 Handsaws at $ 17 and ___ Jackets at $ 12. If the shopper has 10% discount coupon for the Jackets then the total purchase is $73. How much is the missing quantity? [1] 1 [2] 2 [3] 5 [4] 4
8.	Shopper buys 2 T-Shirts at $8 and 2 Mixers at $20. How much returned if the sales tax is 7% and shopper pays $100 ? [1] $38 [2] $43 [3] $40 [4] $37
9.	Shopper buys 2 Handsaws at $16, 2 Handsaws at $14 and 2 Radios at $14. How much is the total purchase? [1] $99 [2] $102 [3] $73 [4] $88
10.	Shopper buys 3 Handsaws at $6, 1 Chair Cushion at $15 and 2 Radios at $16. How much is due if the shopper has 30% discount coupon for the Radios? [1] $63 [2] $59 [3] $47 [4] $55

6. eviF 7. owT 8. eerhT 9. ruoF 10. ruoF

11.	Shopper buys 1 Radio at $20 and 3 Pairs of Shorts at $20. How much is due if the shopper has 10% discount coupon for the Radio?
12.	Shopper buys 1 Pair of Sandals at $15, 2 Hammers at $15 and 3 Pairs of Shorts at $15. How many total items in the shopping cart? [1] 5 [2] 7 [3] 6 [4] 3
13.	Shopper buys ___ T-Shirts at $ 8 and 3 Mixers at $ 20. Total purchase is $76. How much is the missing quantity? [1] 2 [2] 5 [3] 3 [4] 1
14.	Shopper buys 3 Radios at $14, 1 Pair of Flat Sandals at $20 and 2 Backpacks at $14. How much is due if the shopper has 10% discount coupon for the Backpacks?
15.	Shopper buys 1 Hat at $14, 3 Backpacks at $14 and 1 Bakeware Set at $20. How much returned if the sales tax is 10% and shopper pays $100 ?

11. srallod thgiE-ytneveS 12. eerhT 13. enO 14. srallod neveS-ythgiE
15. srallod neetxiS

16.	Shopper buys 3 Polo Shirts at $10, 1 Bakeware Set at $20 and 2 T-Shirts at $8. How much is the difference between the most expensive item and least expensive item? [1] $12 [2] $9 [3] $13 [4] $11
17.	Shopper buys 1 Pair of Shorts at $20, 2 Radios at $___ and 1 Ball at $11. Shopper pays $100 and received the change of $43. How much is the missing unit price?
18.	Shopper buys 3 Polo Shirts at $10, 1 Radio at $14 and 2 Pairs of Sandals at $15. How much change returned if shopper pays $100 ? [1] $25 [2] $29 [3] $26 [4] $30
19.	Shopper buys 1 T-Shirt at $8, 2 Table Linen Sets at $14 and 2 T-Shirts at $11. How much returned if the sales tax is 7% and shopper pays $100 ? [1] $38 [2] $39 [3] $41 [4] $40
20.	Shopper buys 1 Pair of Shorts at $15, 1 Backpack at $14 and 1 Watch at $11. How much change returned if shopper pays $50 ?

16. enO 17. srallod neetrihT 18. eerhT 19. enO 20. srallod neT

21.	Shopper buys 1 Container at $20, 1 Hammer at $16 and 2 Pairs of Sandals at $20. How much returned if the sales tax is 5% and shopper pays $100? [1] $20 [2] $22 [3] $19 [4] $18
22.	Shopper buys 2 Chair Cushions at $ 15 and ___ Pairs of Shorts at $ 20. Total purchase is $90. How much is the missing quantity? [1] 6 [2] 4 [3] 5 [4] 3
23.	Shopper buys 1 Camera at $12, 2 Balls at $11 and 1 Handsaw at $16. How much is the total purchase? [1] $50 [2] $48 [3] $41 [4] $46
24.	Shopper buys 1 Pair of Socks at $7, 2 Hammers at $15 and 1 Pair of Sandals at $20. How much is the difference between the most expensive item and the next most expensive item?
25.	Shopper buys 2 Pairs of Sandals at $15 and 1 Pair of Socks at $7. How much is the total purchase? [1] $37 [2] $42 [3] $31 [4] $38

21. enO 22. ruoF 23. enO 24. srallod eviF 25. enO

26.	Shopper buys 1 Pair of Espadrilles at $20, 1 Backpack at $17 and 1 T-Shirt at $___. Difference between the most expensive item and least expensive item is $12. How much is the missing unit price?
27.	Shopper buys 3 Pairs of Gloves at $8 and 3 Pairs of Jeans at $18. How much is due if the shopper has 20% discount coupon for the Pairs of Gloves? [1] $81 [2] $73 [3] $86 [4] $69
28.	Shopper buys 2 Balls at $11, 1 Pair of Flat Sandals at $20 and 1 Watch at $11. How much is due if the sales tax is 5% ? [1] $56 [2] $47 [3] $63 [4] $66
29.	Shopper buys ____ Pairs of Espadrilles at $ 20 and 3 Pairs of Socks at $ 7. If the shopper has 20% discount coupon for the Pairs of Socks then the total purchase is $57. How much is the missing quantity? [1] 1 [2] 2 [3] 6 [4] 4
30.	Shopper buys 1 Mixer at $20, 3 Pairs of Flat Sandals at $20 and 1 Ball at $11. How much is the difference between the most expensive item and least expensive item? [1] $9 [2] $8 [3] $7 [4] $6

26. srallod thgiE 27. owT 28. enO 29. owT 30. enO

31.	Shopper buys 3 Handsaws at $14 and 3 Glassware Sets at $9. How much is due if the sales tax is 8% ?
32.	Shopper buys 1 Backpack at $17, 2 Balls at $11 and 2 Pairs of Jeans at $20. How much is due if the shopper has 30% discount coupon for the Pairs of Jeans? [1] $55 [2] $64 [3] $75 [4] $67
33.	Shopper buys 1 Pair of Jeans at $18, 3 Glassware Sets at $9 and 2 T-Shirts at $8. How much is due if the sales tax is 10% ? [1] $58 [2] $75 [3] $56 [4] $67
34.	Shopper buys ___ Pairs of Espadrilles at $ 20 and 2 Pairs of Shorts at $ 15. If sales tax is 7% then the total purchase is $75. How much is the missing quantity? [1] 4 [2] 5 [3] 1 [4] 2
35.	Shopper buys 1 Pair of Socks at $7, 1 Pair of Jeans at $20 and 1 Radio at $20. How much is the difference between the most expensive item and least expensive item?

31. srallod eviF-ytneveS 32. ruoF 33. ruoF 34. ruoF 35. srallod neetrihT

36.	Shopper buys 1 Backpack at $____, 1 Watch at $11 and 1 T-Shirt at $8. Shopper pays $50 and received the change of $14. How much is the missing unit price?
37.	Shopper buys 3 Flatware Sets at $11 and 2 Handsaws at $17. How much is the difference between the most expensive item and the next most expensive item? [1] $2 [2] $3 [3] $6 [4] $8
38.	Shopper buys 3 Table Linen Sets at $14 and 2 Handsaws at $6. How many total items in the shopping cart?
39.	Shopper buys 1 Flatware Set at $11, 1 Radio at $13 and 3 Pairs of Socks at $7. How many total items in the shopping cart?
40.	Shopper buys 2 Watches at $ 17 and ____ Chair Cushions at $ 15. If sales tax is 9% then the total purchase is $86. How much is the missing quantity? [1] 5 [2] 1 [3] 3 [4] 4

36. srallod neetneveS 37. eerhT 38. eviF 39. eviF 40. eerhT

41.	Shopper buys 2 Pairs of Socks at $7, 3 Nut Drivers at $8 and 2 Pairs of Shorts at $15. How much returned if the sales tax is 6% and shopper pays $100 ?
42.	Shopper buys 2 Handsaws at $17 and 2 Hammers at $15. How much is the difference between the most expensive item and least expensive item? [1] $3 [2] $6 [3] $2 [4] $1
43.	Shopper buys 2 Pairs of Shorts at $20 and 1 Bakeware Set at $18. How much is due if the sales tax is 10% ? [1] $55 [2] $62 [3] $70 [4] $64
44.	Shopper buys 2 Handsaws at $10 and 3 Mixers at $20. How many total items in the shopping cart? [1] 8 [2] 5 [3] 3 [4] 2
45.	Shopper buys ___ Radios at $ 16, 2 Table Linen Sets at $ 7 and 2 Pairs of Shorts at $ 20. If the shopper has 10% discount coupon for the Radios then the total purchase is $83. How much is the missing quantity? [1] 1 [2] 2 [3] 3 [4] 4

41. srallod thgiE-ytnewT 42. eerhT 43. ruoF 44. owT 45. owT

46.	Shopper buys 2 Pairs of Espadrilles at $20, 3 Pairs of Shorts at $15 and 1 Pair of Gloves at $____. Shopper pays $100 and received the change of $7. How much is the missing unit price? [1] $7 [2] $6 [3] $8 [4] $10
47.	Shopper buys 2 Radios at $14, 2 Mobile Phones at $20 and 1 Polo Shirt at $10. How much is the total purchase?
48.	Shopper buys 3 Pairs of Gloves at $8, 1 Nut Driver at $8 and 2 Radios at $16. How much is due if the shopper has 10% discount coupon for the Nut Driver? [1] $69 [2] $57 [3] $51 [4] $63
49.	Shopper buys 1 Pair of Gloves at $8, 1 Hammer at $16 and 3 Containers at $20. How much is due if the sales tax is 5% ?
50.	Shopper buys 2 Balls at $ 19 and ____ Balls at $ 11. If the shopper has 30% discount coupon for the Balls then the total purchase is $49. How much is the missing quantity? [1] 1 [2] 2 [3] 4 [4] 3

46. eerhT   47. srallod thgiE-ytneveS   48. ruoF   49. srallod thgiE-ythgiE
50. owT

51.	Shopper buys 3 Jackets at $___, 1 Top at $18 and 2 Balls at $11. Total purchase is $76. How much is the missing unit price? [1] $11 [2] $12 [3] $9 [4] $10
52.	Shopper buys ___ Watches at $ 11, 3 Radios at $ 7 and 3 T-Shirts at $ 8. $16 returned if the sales tax is 8% and shopper pays $100. How much is the missing quantity?
53.	Shopper buys 2 Polo Shirts at $10, 2 Table Linen Sets at $14 and 1 Pair of Jeans at $18. How much is the difference between the most expensive item and the next most expensive item?
54.	Shopper buys 2 Pairs of Pants at $20, 1 Handsaw at $14 and 1 T-Shirt at $8. How many total items in the shopping cart? [1] 5 [2] 4 [3] 7 [4] 3
55.	Shopper buys 2 Pairs of Pants at $17 and 3 Bakeware Sets at $18. How much returned if the sales tax is 8% and shopper pays $100? [1] $3 [2] $6 [3] $5 [4] $4

51. owT 52. eerhT 53. srallod ruoF 54. owT 55. eerhT

56.	Shopper buys 1 Ball at $19 and 1 Bakeware Set at $18. How much is due if the shopper has 20% discount coupon for the Bakeware Set? [1] $33 [2] $32 [3] $31 [4] $26
57.	Shopper buys 1 Blazer at $17, 2 Pairs of Sandals at $15 and 2 T-Shirts at $8. How much is the difference between the most expensive item and least expensive item? [1] $9 [2] $7 [3] $11 [4] $8
58.	Shopper buys 2 Pairs of Jeans at $20, 1 Ball at $11 and 1 Pair of Shorts at $20. How much is due if the sales tax is 10% ?
59.	Shopper buys ____ Pairs of Flat Sandals at $ 20 and 3 T-Shirts at $ 6. Total purchase is $78. How much is the missing quantity?
60.	Shopper buys 1 Handsaw at $17, 2 Polo Shirts at $10 and 1 Pair of Socks at $7. How much is the difference between the most expensive item and least expensive item? [1] $8 [2] $11 [3] $10 [4] $9

56. enO 57. enO 58. srallod thgiE-ytneveS 59. eerhT 60. eerhT

61.	Shopper buys 3 Watches at $11 and 3 Pairs of Shorts at $20. How much change returned if shopper pays $100 ? [1] $6 [2] $9 [3] $7 [4] $5
62.	Shopper buys 3 Cameras at $___ and 3 Table Linen Sets at $14. If the shopper has 20% discount coupon for the Table Linen Sets then the total purchase is $70. How much is the missing unit price?
63.	Shopper buys 2 Cameras at $12, 1 Handsaw at $16 and 1 Blazer at $17. How much is due if the shopper has 30% discount coupon for the Handsaw?
64.	Shopper buys 1 Blazer at $17, 2 Containers at $___ and 1 Handsaw at $6. If sales tax is 9% then the total purchase is $69. How much is the missing unit price? [1] $20 [2] $22 [3] $18 [4] $16
65.	Shopper buys 3 Hammers at $16 and 2 T-Shirts at $6. How much is the difference between the most expensive item and the next most expensive item?

61. eerhT 62. srallod evlewT 63. srallod owT-ytfiF 64. enO 65. srallod neT

66.	Shopper buys 1 T-Shirt at $8, 1 Bakeware Set at $20 and 3 Radios at $14. How much is the difference between the most expensive item and the next most expensive item? [1] $6 [2] $5 [3] $3 [4] $7
67.	Shopper buys 3 Backpacks at $14, 1 Radio at $7 and 2 Pairs of Espadrilles at $___. Shopper pays $100 and received the change of $17. How much is the missing unit price?
68.	Shopper buys 3 Mixers at $20 and 1 Glassware Set at $9. How much is the total purchase?
69.	Shopper buys 3 Table Linen Sets at $7, 1 Nut Driver at $8 and 2 Mobile Phones at $___. If sales tax is 7% then the total purchase is $74. How much is the missing unit price?
70.	Shopper buys ___ Radios at $ 7, 2 Backpacks at $ 17 and 3 Balls at $ 11. 7 items in the basket. How much is the missing quantity?

66. enO 67. srallod neetneveS 68. srallod eniN-ytxiS 69. srallod -ytnewT 70. owT

| 71. | Shopper buys 3 Skirts at $17 and 2 Containers at $20. How much is the difference between the most expensive item and the next most expensive item? |

| 72. | Shopper buys 2 Pairs of Sandals at $15, 1 Pair of Espadrilles at $20 and 1 Handsaw at $17. How much is the total purchase? [1] $67 [2] $60 [3] $76 [4] $63 |

| 73. | Shopper buys 1 Pair of Sandals at $20, 2 Balls at $11 and 3 Glassware Sets at $9. How much is the total purchase? |

| 74. | Shopper buys 1 Pair of Espadrilles at $17, 2 Nut Drivers at $8 and 1 Skirt at $17. How much is the total purchase? |

| 75. | Shopper buys 3 Balls at $11, 2 Pairs of Jeans at $20 and 1 Pair of Sandals at $15. How much is the difference between the most expensive item and least expensive item? |

71. srallod eerhT 72. enO 73. srallod eniN-ytxiS 74. srallod -ytfiF
75. srallod eniN

76.	Shopper buys 2 Backpacks at $14 and 3 Pairs of Pants at $20. How much returned if the sales tax is 8% and shopper pays $100 ?
77.	Shopper buys 2 Pairs of Shorts at $20, 1 Pair of Gloves at $8 and 1 Radio at $___. Difference between the most expensive item and the next most expensive item is $4. How much is the missing unit price?
78.	Shopper buys 2 Hats at $14, 1 Watch at $11 and 2 Backpacks at $11. How much is due if the sales tax is 9% ? [1] $72 [2] $52 [3] $56 [4] $66
79.	Shopper buys 1 Pair of Shorts at $15, 2 Balls at $11 and 3 Jackets at $12. How much returned if the sales tax is 8% and shopper pays $100 ? [1] $18 [2] $22 [3] $17 [4] $21
80.	Shopper buys 3 Pairs of Sandals at $15 and 2 Handsaws at $10. How much is due if the shopper has 10% discount coupon for the Pairs of Sandals? [1] $61 [2] $56 [3] $51 [4] $57

76. srallod eviF 77. srallod neetxiS 78. ruoF 79. ruoF 80. enO

81.
Shopper buys 1 Chair Cushion at $15, 2 Pairs of Shorts at $15 and 2 Handsaws at $6. How much is the total purchase?
[1] $53 [2] $57 [3] $45 [4] $47

82.
Shopper buys 1 Handsaw at $10, 3 Radios at $14 and 1 Pair of Sandals at $20. How much is the difference between the most expensive item and least expensive item?

83.
Shopper buys 2 Balls at $ 19 and ___ Radios at $ 13. 5 items in the basket. How much is the missing quantity?   [1] 4
[2] 6 [3] 3 [4] 2

84.
Shopper buys 3 Backpacks at $14, 2 Cameras at $12 and 1 Pair of Flat Sandals at $20. How many total items in the shopping cart?
[1] 8 [2] 7 [3] 9 [4] 6

85.
Shopper buys 1 Handsaw at $14 and 3 Radios at $___. If the shopper has 30% discount coupon for the Handsaw then the total purchase is $70. How much is the missing unit price?

81. owT 82. srallod neT 83. eerhT 84. ruoF 85. srallod -ytnewT

86.	Shopper buys 3 Polo Shirts at $10 and 3 Backpacks at $17. How much is due if the shopper has 10% discount coupon for the Backpacks? [1] $66 [2] $83 [3] $65 [4] $76
87.	Shopper buys 3 Radios at $13, 2 Polo Shirts at $10 and 1 Backpack at $11. How much change returned if shopper pays $100 ? [1] $32 [2] $25 [3] $24 [4] $30
88.	Shopper buys 3 Pairs of Shorts at $15 and 2 Polo Shirts at $10. How much returned if the sales tax is 8% and shopper pays $100 ?
89.	Shopper buys 1 Ball at $11, 3 Hammers at $15 and 2 Radios at $7. How much is due if the shopper has 10% discount coupon for the Ball?
90.	Shopper buys 3 Pairs of Gloves at $8 and 2 Pairs of Jeans at $___. Total purchase is $64. How much is the missing unit price?

86. ruoF 87. ruoF 88. srallod -ytrihT 89. srallod eniN-ytxiS 90. srallod - ytnewT

STATUE BY THE SOUTH EAST CORNER OF CENTRAL PARK IN NEW YORK CITY

CHAPTER 4: K-4 MATH EXERCISES

www.ingramcontent.com/pod-product-compliance
Lightning Source LLC
Chambersburg PA
CBHW081430170526
45166CB00008B/2148